Hayner Public Library District - Alton
P9-EMF-898

ANIMAL AND PLANT
Anatomy

VOLUME CONSULTANTS

• Barbara Abraham, *Hampton University, VA* • Amy-Jane Beer, *Natural history writer and consultant*
• Thomas Kunz, *Boston University, MA* • Chris Mattison, *Natural history writer and researcher*
• Richard Mooi, *California Academy of Sciences, San Francisco, CA*
• Adrian Seymour, *Bristol University, England*

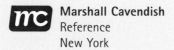

3

Fruit bat – Grizzly bear

Marshall Cavendish
Reference
New York

Contributors
Roger Avery; Richard Beatty; Amy-Jane Beer; Erica Bower; Trevor Day; Erin Dolan; Bridget Giles; Natalie Goldstein; Tim Harris; Christer Hogstrand; Rob Houston; John Jackson; Tom Jackson; James Martin; Chris Mattison; Katie Parsons; Ray Perrins; Kieran Pitts; Adrian Seymour; Steven Swaby; John Woodward.

Consultants
Barbara Abraham, Hampton University, VA; Glen Alm, University of Guelph, Ontario, Canada; Roger Avery, Bristol University, England; Amy-Jane Beer, University of London, England; Deborah Bodolus, East Stroudsburg University, PA; Allan Bornstein, Southeast Missouri State University, MO; Erica Bower, University of London, England; John Cline, University of Guelph, Ontario, Canada; Trevor Day, University of Bath, England; John Friel, Cornell University, NY; Valerius Geist, University of Calgary, Alberta, Canada; John Gittleman, University of Virginia, VA; Tom Jenner, Academia Británica Cuscatleca, El Salvador; Bill Kleindl, University of Washington, Seattle, WA; Thomas Kunz, Boston University, MA; Alan Leonard, Florida Institute of Technology, FL; Sally-Anne Mahoney, Bristol University, England; Chris Mattison; Andrew Methven, Eastern Illinois University, IL; Graham Mitchell, King's College, London, England; Richard Mooi, California Academy of Sciences, San Francisco, CA; Ray Perrins, Bristol University, England; Kieran Pitts, Bristol University, England; Adrian Seymour, Bristol University, England; David Spooner, University of Wisconsin, WI; John Stewart, Natural History Museum, London, England; Erik Terdal, Northeastern State University, Broken Arrow, OK; Phil Whitfield, King's College, University of London, England.

Marshall Cavendish
99 White Plains Road
Tarrytown, NY 10591–9001

www.marshallcavendish.us

Library of Congress Cataloging-in-Publication Data
Animal and plant anatomy.
p. cm.
ISBN-13: 978-0-7614-7662-7 (set: alk. paper)
ISBN-10: 0-7614-7662-8 (set: alk. paper)
ISBN-13: 978-0-7614-7666-5 (vol. 3)
ISBN-10: 0-7614-7666-0 vol. 3 (vol. 3)
1. Anatomy. 2. Plant anatomy. I. Marshall Cavendish Corporation. II.
Title.

QL805.A55 2006
571.3--dc22

2005053193

Printed in China
09 08 07 06 1 2 3 4 5

Marshall Cavendish
Editor: Joyce Tavolacci
Editorial Director: Paul Bernabeo
Production Manager: Mike Esposito

The Brown Reference Group plc
Project Editor: Tim Harris
Deputy Editor: Paul Thompson
Subeditors: Jolyon Goddard, Amy-Jane Beer, Susan Watts
Designers: Bob Burroughs, Stefan Morris
Picture Researchers: Susy Forbes, Laila Torsun
Indexer: Kay Ollerenshaw
Illustrators: The Art Agency, Mick Loates, Michael Woods
Managing Editor: Bridget Giles

Photographic credits, volume 3

Cover illustration is a composite of anatomical illustrations found within this encyclopedia.

Ardea: Brakefield, Tom 313, 321; Fink, Kenneth W. 349; Gohier, Francios 375, 408; Parer, D. & Parer-Cook, E. 355; Rabinowitz, Neil 391; Warren, Adrian 401; **FLPA:** Hosking, Stephen 348, Minden Pictures 390; **John Foxx Images:** 363; **NaturePL:** Brock, Richard 331; Douwma, Georgette 345; Foott, Jeff 394; Maynerd, Hugh 293; Nill, Dietmar 300, 302; Oxford, Pete 326; Pusser, Tod 393; Savigny, Francios 400; Shah, Anup 328; Vezo, Tom 333; White, Doc 379; **NHPA:** Griffith, Ken 305; Jones, B. & Shimlock M. 343; Wendler, Martin 397; **NOAA:** Roletto, Jan 378; **OSF:** 353; **Photodisc:** 411; Gritzt, James 384; Hollingsworth, Jack 357; **Photos.com:** 361; 371; 373; 427; **Still Pictures:** Arnold, Peter 425; Kazlowski, Steven 429; Sauvenet, Jany 324; **Thomas Eisner:** 340; **Topham:** 366.

Contents

Fruit bat

ORDER: Chiroptera SUBORDER: Megachiroptera
FAMILY: Pteropodidae

Bats are the only mammals to have evolved powered flight. There are 1,116 species of bats known to biologists, making up one-fifth of all mammal species. Most bats are tiny and emerge at dusk to dart after night-flying insects. By contrast, some bats in the tropical forests of Africa, Asia, and Australasia grow far larger and use their huge eyes to forage for fruit in the treetops. They are the Old World fruit bats, and some of these are called flying foxes.

Anatomy and taxonomy

Biologists categorize all organisms into groups based on anatomical and genetic evidence. All the Old World fruit bats belong to a single family, the Pteropodidae.

- **Animals** Animals are multicellular (many-celled) organisms that rely on other organisms, usually by feeding on them, to obtain energy and nutrients.

- **Chordates** At some time in its life cycle a chordate has a stiff, dorsal (back) supporting rod called the notochord that runs along most of the length of its body.

- **Vertebrates** Vertebrates are chordates whose notochord transforms into a backbone, or spinal column, during embryonic development. The backbone composes a chain of small elements called vertebrae, made of cartilage or bone. Vertebrates also have a braincase, or cranium.

- **Mammals** Mammals are vertebrates with mammary glands that secrete milk, a nutritious substance that feeds the growing young. All mammals have hair, and their lower jaw consists of a single bone.

- **Placental mammals** Placental mammals, or eutherians, nourish their developing young for an extended period inside the uterus through an organ called a placenta. Placental mammals include all living mammals except marsupials and monotremes (platypuses and echidnas).

- **Bats** Bats are placental mammals that travel by powered flight. They fly by flapping wings of skin stretched between the long hand and finger bones. With a few exceptions, bats are active only at night. The hind legs of bats are uniquely adapted for hanging. Over the course of evolution they have rotated 180 degrees relative to the pelvis and thus the knees appear to flex forward, not backward like those of other mammals.

- **Microbats** These bats avoid obstacles and hunt prey in complete darkness using echolocation. They form sound images by sending out pulses of high-frequency ultrasound

▶ *All Old World fruit bats belong to a single family, the Pteropodidae (not shown). The relationship of microbats and Old World fruit bats is under debate; controversy rages over whether the bat groups form one distinct order, as shown here, or whether fruit bats are actually more closely related to primates such as humans.*

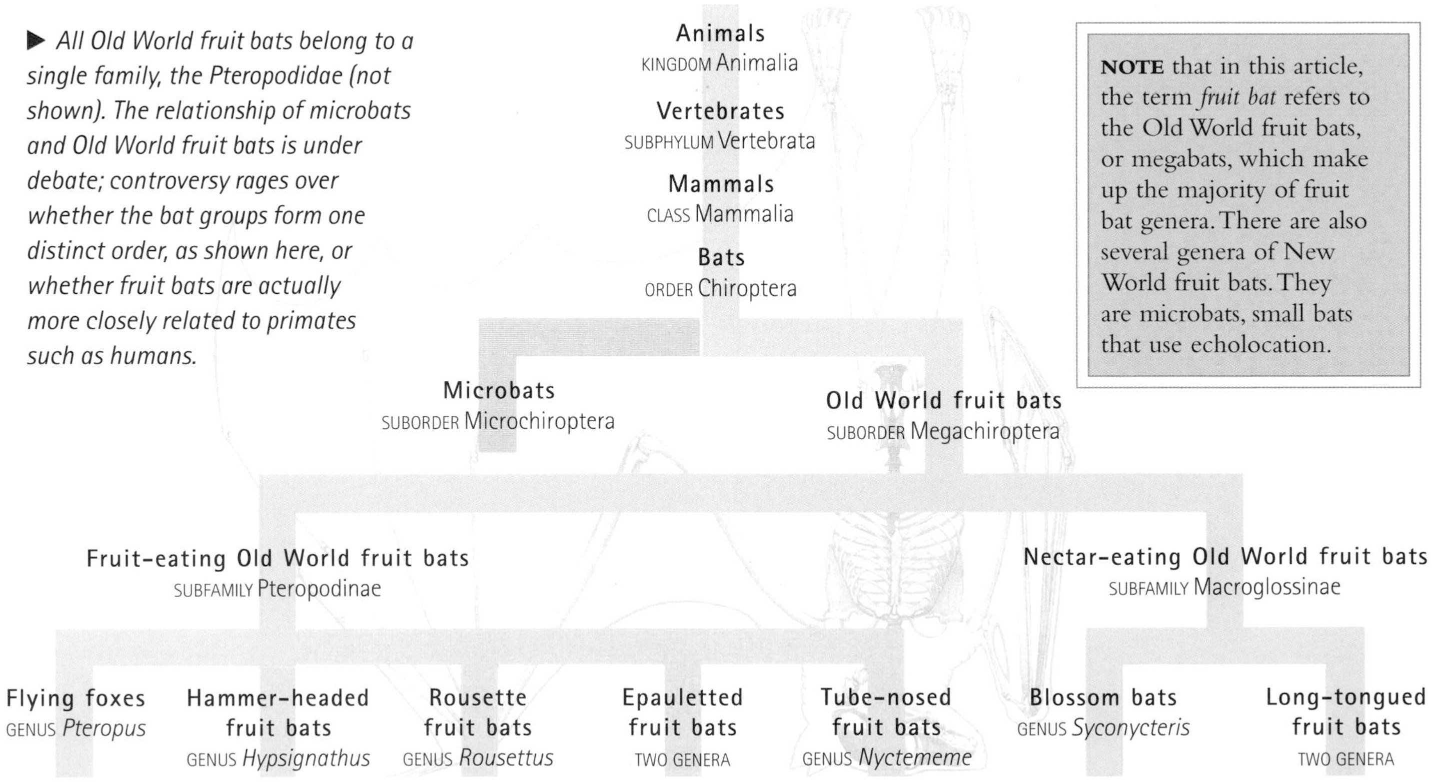

NOTE that in this article, the term *fruit bat* refers to the Old World fruit bats, or megabats, which make up the majority of fruit bat genera. There are also several genera of New World fruit bats. They are microbats, small bats that use echolocation.

and interpreting the echoes from surrounding objects. Microbats have complex external ears with a pinna (ear funnel) and tragus (a small rod or flap at the pinna entrance); there may also be a variety of ornaments (noseleaves, horseshoes, or flaps) on the face that shape the outgoing sound beam. The cochlea in the inner ear is large, and the auditory pathway of the brain, particularly the inferior colliculus, is enlarged. Most microbats weigh less than 0.5 ounce (15 g). The order Microchiroptera includes the phyllostomids, or New World fruit bats, which live in tropical South and Central America.

• Old World fruit bats Other than the rousette fruit bats, Old World fruit bats (also called megabats or megachiropterans) do not echolocate. Instead, they have large, forward-facing eyes with visual pathways from each eye to both halves of the brain. This gives them binocular vision, with good distance judgment, like primates. Fruit bats feed on fruit, nectar, pollen, or all three; some also feed on leaves. They have a claw on the second finger as well as a thumb. Fruit bats have no tail and, unlike microbats, have very little wing membrane between the legs. Fruit bats' ears are simple and lack a tragus on the pinna, or external ear, and they do not have an enlarged cochlea or inferior colliculus. Some Old World fruit bats are large, with a wingspan of up to 6 feet (1.8 m). However, most species are much smaller, and many are as small as microbats.

• Nectar-eating Old World fruit bats Nectar- and pollen-feeding fruit bats, or macroglossines, have a long, tubelike snout that reaches the bases of long flowers. They have a long tongue with papillae (fleshy projections) that soak up nectar but have few teeth.

• Fruit-eating Old World fruit bats Most fruit bats are fruit eaters that lack the specific adaptations of nectar-eating bats. Also called pteropodine bats, they have a doglike face, although some look more like monkeys with wings. Their short jaws give a powerful bite, and their cheek teeth can crush the toughest fruit. Some have bizarre nostrils extending from the snout. Pteropodines range from the large flying foxes to the tiny pygmy fruit bat, which is less than 3 inches (7 cm) long.

▲ *A roosting gray-headed fruit bat. These large flying foxes live throughout eastern Australia and feed on fruit, especially figs.*

• Flying foxes The 59 Old World fruit bats in the genus *Pteropus* are called flying foxes. They differ in size, in color (from black to brown, red, or gray), and in social behavior, but they have a similar body shape, with a doglike snout, large, more or less forward-facing eyes, and broad wings.

FEATURED SYSTEMS

EXTERNAL ANATOMY A fruit bat's furry body takes flight on broad, membranous wings supported by immensely long hand and finger bones. *See pages 294–297.*

SKELETAL SYSTEM Bats can fly and hang upside down, thanks to unusual skeletal features. *See pages 298–300.*

MUSCULAR SYSTEM Bats' largest muscles power their beating wings. They also have specialized muscles that control their wing shape. *See pages 301–302.*

NERVOUS SYSTEM Fruit bats have a large brain, which allows them to have complex social lives. Their eyes allow superb night vision, and their strong sense of smell helps them locate ripe fruit. *See pages 303–305.*

CIRCULATORY AND RESPIRATORY SYSTEMS Powered flight demands a great deal of energy. Fruit bats need a powerful and efficient heart and lungs to supply energy to the body. *See pages 306–307.*

DIGESTIVE AND EXCRETORY SYSTEMS Fruit bats deal with a low-protein, high-energy diet by eating large amounts of food and absorbing it rapidly. *See pages 308–309.*

REPRODUCTIVE SYSTEM Some species of fruit bats have specialized anatomical features that attract mates, such as hair tufts and scent glands. Young are born well-developed but cannot fly until they approach adult size. *See pages 310–311.*

External anatomy

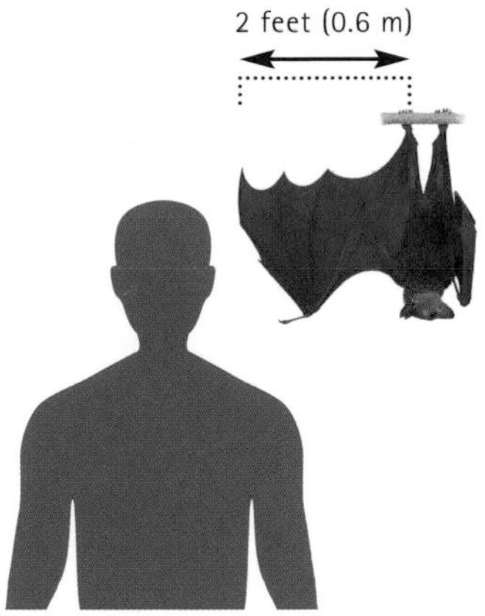
2 feet (0.6 m)

CONNECTIONS

COMPARE the shape of a fruit bat's wings with those of a ***HUMMINGBIRD***. Most of a fruit bat's wing consists of membrane stretched between the extended digits. However, the digits of a hummingbird's wing are short relative to the length of the wing and do not support the flight surface.

Bats have characteristically mammalian teeth, jaws, and skulls. They also have typical mammalian skin. The trunk and head have a dense coating of fur, some 0.4 inch (1 cm) thick. And female bats have another important mammal-defining feature: mammary glands, which secrete nourishing milk through the nipples to feed the young.

Wing shape and structure

Bats' wings dominate their appearance. The wings are composed of a thin double-layered membrane of skin called a patagium. Muscles, connective tissue, nerves, and blood vessels are sandwiched between the layers. The wing membrane is stretched between the bat's body and arms and between the long fingers to the legs. It also extends a little in front of the arm.

The arm, hand, and finger bones form a collapsible support structure like the spokes of an umbrella. When the wing is not extended, it is folded up, with the wing membrane crumpling and tucked away safely. Oily secretions from glands keep the skin of the wing moist and supple. The surface of the wing membrane is covered with microscopic

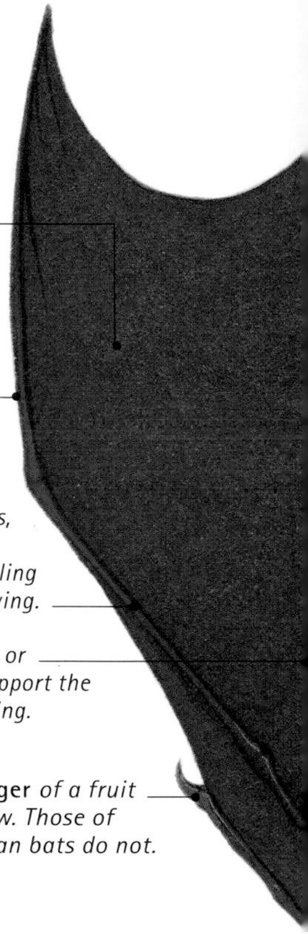
The **chiropatagium** *lies between the finger bones. It is the part of the wing that provides most of the forward thrust in flight.*
These bones are raised above the wing surface. This helps create a boundary layer around the wing that reduces drag.
The finger bones, or **phalanges**, *support the trailing section of the wing.*
The hand bones, or **metacarpals**, *support the middle of the wing.*
The **second finger** *of a fruit bat bears a claw. Those of microchiropteran bats do not.*

COMPARATIVE ANATOMY

Old and New Worlds

The appearance of a fruit bat's face is dominated by the animal's reliance on vision. It has relatively small, simple external ears, large eyes, and a snout not unlike that of a dog or lemur. Its face contrasts with that of a microbat, which uses echolocation to find its way about. New World fruit bats are microbats. They have large, complex external ears with tragi, small eyes, and often an elaborate projection called a nose-leaf on their snout. Their facial appearance is dominated by the requirement for echolocation, just as that of Old World fruit bats is influenced by the need for sharp eyesight at night.

▼ Gray-headed fruit bat
Old World fruit bats do not echolocate but rely instead on sight; they have very large eyes with the capacity for binocular vision.

▲ Leaf-nosed bat
This New World fruit bat's face bears nasal adaptations for directing echolocation pulses.

*Over the course of evolution, bat **legs** have rotated by 180 degrees relative to the pelvis. The bones still flex at the knee in the same direction, though; the net effect is that the knees appear to flex in the opposite direction from all other animals.*

*Other than those of the first two fingers, the **fingertips** have no trace of nails, claws, or pads.*

*The **plagiopatagium** is the part of the wing between the fifth finger and the trunk. It produces most of the lift during flight.*

*All five **toes** face the same way, with the claws curving forward. This is unusual in mammals.*

*Fruit bats have only a small flap of wing membrane attached to each leg, the **interfemoral membrane**.*

*The **fur** of fruit bats is usually black or brown, though it can also be gray, white, red, or orange. A gray-headed fruit bat has dark brown wings and trunk, a yellowish collar (or mantle), and a silver-gray head.*

*The **propatagium** is the part of the wing in front of the arm bones.*

*The **eyes** are very large. This maximizes the amount of dim light that can be gathered for detection.*

Fruit bats generally do not echolocate, so there is no need for elaborate ornaments on the nostrils to direct sound beams.

*The **thumb** is large, powerful, and mobile in fruit bats. It is used to support the animal as it clambers, and also to hold food.*

*The **pinna**, or external ear, is simple and does not have a tragus as those of microchiropteran bats do.*

hairs. Biologists think that the hairs are mechanical detectors that send information to the brain about the flow of air over the wing. The hairs, together with irregularities in the wing surface created by the knobbly bones beneath, help form a "boundary" layer of weakly turbulent air around the wing in flight. This boundary layer allows the bat to slip through the air with little drag (drag is the force caused by the resistance of air to the movement of objects through it).

Thumb power

Although most of a bat's fingers are employed in holding out and collapsing the wing, the thumb is free for other tasks. The thumbs of fruit bats are larger and more powerful than those of microbats. Fruit bats use their thumbs

▲ Gray-headed fruit bat
Wings are essential not just for flight; they also act as body temperature regulators and rain shields and are used for cradling young.

COMPARATIVE ANATOMY

Bat tails

Most microbats have a tail supported by a number of caudal (tail) vertebral bones. A membrane called the uropatagium stretches between the legs and the tail. These bats also use their tail membrane for lift during flight, and it becomes especially useful as an air brake, helping the animal maneuver or scoop up prey. Microbats are extremely agile, thanks in part to their tail membrane, the size of which may reflect the animal's feeding strategy. Fruit bats do not catch prey on the wing and do not maneuver in dense undergrowth while flying—most are just too big. Fruit bats fly for the most part in straight, level commuting journeys from roost to feeding grounds. They have no tail and only small membranes between their legs. This reduces drag as they fly and makes level flight more efficient. It also frees the back legs and allows a fruit bat greater clambering ability.

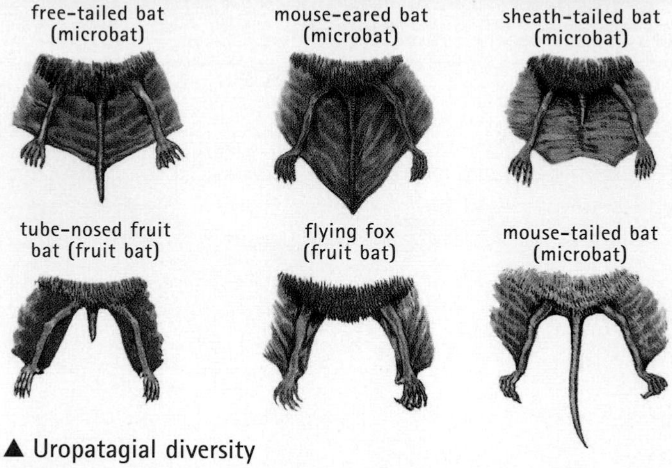

▲ **Uropatagial diversity**

Mouse-tailed bats have a long tail that resembles that of other microbats. However, they lack a uropatagium linking the tail to the legs.

IN FOCUS

Powered flight

Wings work by creating lift, an upward force that overcomes the downward force of gravity on the animal's body mass. A wing acts as an airfoil as it moves through air. It forces air moving over its upper surface to speed up relative to air moving underneath. As it speeds up, the air is stretched out over the airfoil's curved shape, so the air becomes less dense than the air below. The dense air beneath the airfoil exerts the upward lift force as it tries to rush to fill the gap. Flying animals must produce another force, thrust, to overcome drag, which is the resistance of air to movement through it. Thrust is provided by the flapping of the outer parts of the wings. The inner wing sections produce lift whether the wings are moving up or down.

▼ **Wing movements**

The path of the wing tips is shown by the dotted line. During the downstroke, the wing extends fully and beats down and forward. The upstroke sees the wing move back and up, with the outer part of the wing retracted.

The wings beat forward during the downstroke and back during the upstroke.

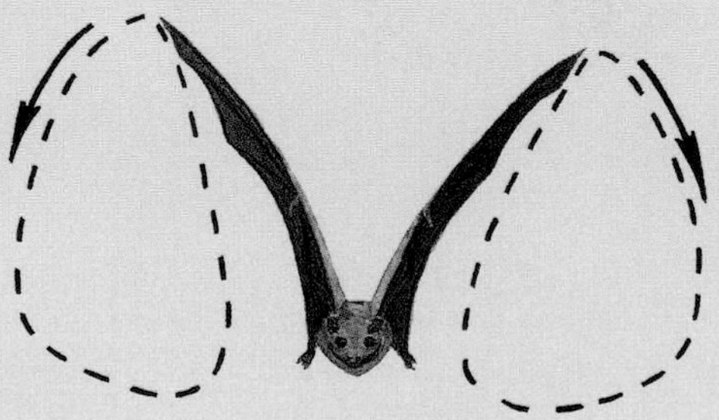

The wings are fully extended during the downstroke, but are swept in at the wrist during the upstroke.

for grasping and pulling at fruit and for climbing through the treetops. The bats are able to hang head upward from one of their thumb claws while urinating or defecating.

Unlike microbats, fruit bats also have a claw on their second finger. This is used in combination with the thumb for grasping objects and for climbing.

Cooling system

Wings serve many purposes besides flight. They provide a huge surface across which bats can quickly lose excess heat. By shunting blood from the body core to the wings, a bat can swiftly transfer heat from its body to the surroundings. The wings act as radiators. At every part of the wing, hot blood is only fractions of an inch from the surface, so the blood cools rapidly by conduction and radiation to the surrounding air.

Many flying foxes encourage cooling by salivating on their wings and by fanning them while hanging at their roost. The saliva evaporates and cools the bat; heat energy is drawn in to evaporate the saliva.

Eyes forward

Another prominent feature is the eyes. They are large enough to produce sharp images even in dim nocturnal light. Fruit bats get around and find their food mainly by vision, even on the darkest night. The eyeball of a large flying fox can be up to 0.4 inch (1 cm) across. More unusually, fruit bats' eyes face almost directly forward, while those of most other mammals face sideways.

The few mammals whose eyes face forward are mainly primates such as monkeys, apes, and humans, which all have eyes facing directly forward. Fruit bats' eyes face in slightly different directions; they diverge by an angle of around 20 degrees. Even so, a large proportion of what they see (their visual field) is seen by both eyes at once. This arrangement allows binocular vision, which permits accurate perception of distances. This is particularly important for animals that reach and grasp.

IN FOCUS

Flying fox aerodynamics

Fruit bats, particularly the large flying foxes, need strength, endurance, and economy from their wings, rather than maneuverability. The larger species commute many miles each night to their feeding grounds and migrate over long distances. Gray-headed fruit bats may migrate 450 miles (750 km) or more. They have powerful wing muscles, and broad wings with a large surface area that maximizes lift. Unlike other animals with broad wings, however, the larger fruit bats are not maneuverable or agile. Broad wings prevent fruit bats from flying fast, because the wings create a lot of drag. Nevertheless, flying foxes' flight is fairly economical. Their wings provide great lift, allowing female flying foxes to carry young pups half their own weight.

EVOLUTION

Gliding lemurs and flying bats

The closest living relatives of bats are probably the flying lemurs of the order Dermaptera. They are neither lemurs nor primates. In some ways, however, flying lemurs may resemble a close ancestor of bats. Flying lemurs, like other gliding mammals such as sugar gliders and flying squirrels, have a flight membrane that forms a lift-generating airfoil shape.

Flying lemurs support their flight membrane with arm, leg, hand, and finger bones. However, they can only glide; they cannot maintain powered flight. They glide from tree to tree, but cannot gain or maintain height during airborne journeys. Both bats' and flying lemurs' wings produce lift; the crucial difference is that bats can flap their wings to produce another force, thrust, which allows them to power and control their flight consistently. The evolution of flight in bats is a mystery; the earliest fossil bats are fully equipped for powered flight, and are anatomically almost identical to modern microchiropterans. However, most experts agree that before bats evolved powered flight, they must have gone through a stage of gliding from tree to tree, much as flying lemurs do today.

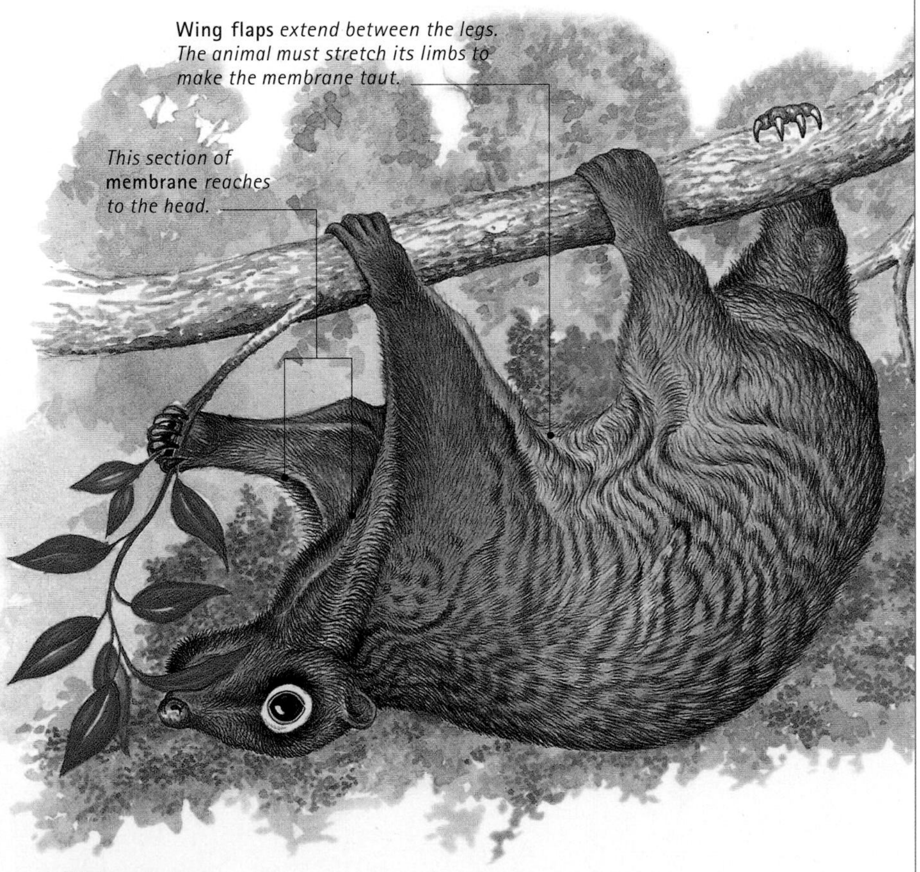

Wing flaps extend between the legs. The animal must stretch its limbs to make the membrane taut.
This section of membrane reaches to the head.

▲ **Flying lemur**
Flaps of skin provide a lifting surface for a flying lemur. Although fossils of bat ancestors have yet to be discovered, they probably glided in a similar way.

Skeletal system

CONNECTIONS

COMPARE the pivot points of a fruit bat's wings with those of an ***EAGLE.*** The eagle's wing bones pivot at the shoulder; those of a bat pivot at the joint between clavicle and sternum.

COMPARE the adaptations of the hand bones of a fruit bat with those of a ***COELACANTH*** and a ***HUMAN.***

A bat's skeleton is a framework of rigid plates and rods within the body. The skeleton provides protection for vital soft parts, support for the body, and leverage, in conjunction with muscles, whenever the animal moves. A bat's light but strong skeleton supports its wings and allows them to beat in ways not found in any other animal group.

The wing bones

The wing bones are responsible for keeping the wings rigid, preventing them from being distorted by the powerful forces generated during flight. The wings contain the same bones as in the front limbs of other mammals. All the usual bones are there—humerus, ulna, and radius, hand bones (metacarpals), and finger bones (phalanges)—but most are very long, particularly the metacarpals and phalanges, and support the wing membrane. Only the thumb bones remain short; the thumb does not support the wing, but it is mobile for grasping things.

The elbow and wrist joints can move only horizontally. The bones move in grooved joints. Projections on each bone interlock with projections on the next, allowing the bones to rotate in one axis but not in any other. These restrictions make the wing stronger and more resistant to deformation. Knobs on the joints also prevent overextension (excessive straightening) of the wing. On the elbow there is a bone unique to bats called the ulnar sesamoid, which prevents overextension.

A bat wing is prone to being flipped by air pressure during the wing beat, just like an umbrella being turned inside out by the wind. The fifth finger points directly backward from the wrist and is extremely rigid at the wrist to help prevent flipping.

▶ Gray-headed fruit bat
Note the unique calcar projection at the bat's heel. It supports the flight membrane close to the leg.

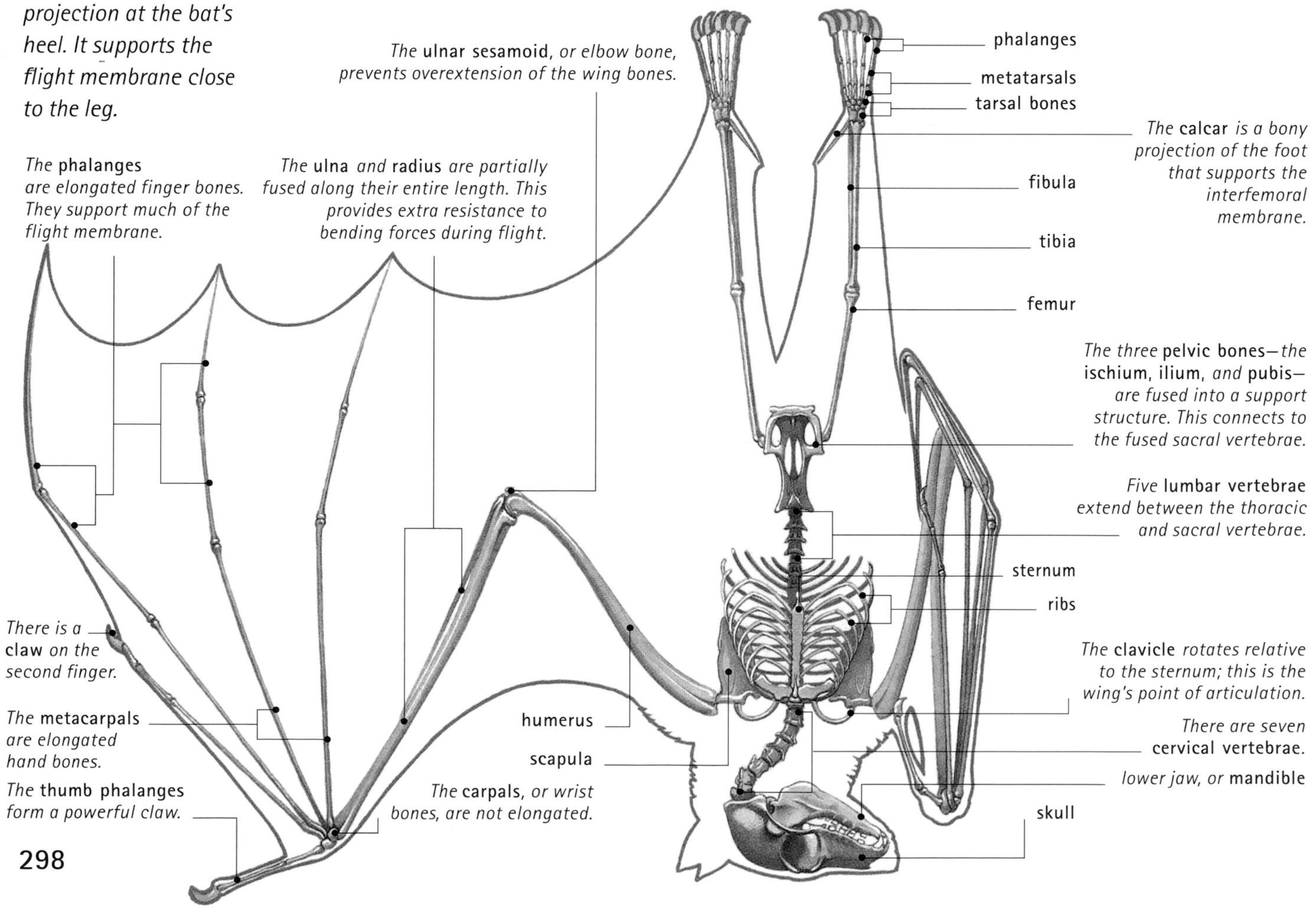

The ulnar sesamoid, or elbow bone, prevents overextension of the wing bones.
phalanges
metatarsals
tarsal bones
The calcar is a bony projection of the foot that supports the interfemoral membrane.
The phalanges are elongated finger bones. They support much of the flight membrane.
The ulna and radius are partially fused along their entire length. This provides extra resistance to bending forces during flight.
fibula
tibia
femur
The three pelvic bones—the ischium, ilium, and pubis—are fused into a support structure. This connects to the fused sacral vertebrae.
Five lumbar vertebrae extend between the thoracic and sacral vertebrae.
sternum
ribs
There is a claw on the second finger.
The clavicle rotates relative to the sternum; this is the wing's point of articulation.
The metacarpals are elongated hand bones.
humerus
There are seven cervical vertebrae.
scapula
lower jaw, or mandible
The thumb phalanges form a powerful claw.
The carpals, or wrist bones, are not elongated.
skull

COMPARATIVE ANATOMY

Transformed limbs

Natural selection has modified the basic vertebrate forelimb into a wing in three completely different ways at different times in evolutionary history. The three groups of animals that resulted were birds, bats, and extinct flying reptiles called pterosaurs. The pterosaur wing was supported only along its leading edge by an immensely long fourth digit. It was stiffened by rigid fibers within the leathery wing tissue. Bird wings are based on the bones of the arm. They also support only the leading edge of the wing. The wing holds its shape owing to the stiffness of the shafts of the flight feathers.

Bat wings have an altogether different structure. Their wings are made of a thin double-layered membrane of skin, with muscles, connective tissue, nerves, and blood vessels in between. The membrane is supported and held rigid by the elongated hand and finger bones.

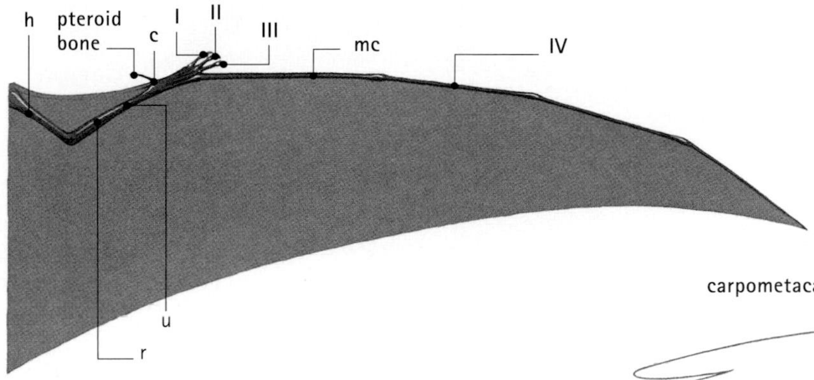

◀ **Pterosaur wing**

The wing membrane was supported on its leading edge by the fourth digit. Pterosaurs had a unique bone, the pteroid, which supported the membrane in front of the lower arm bones.

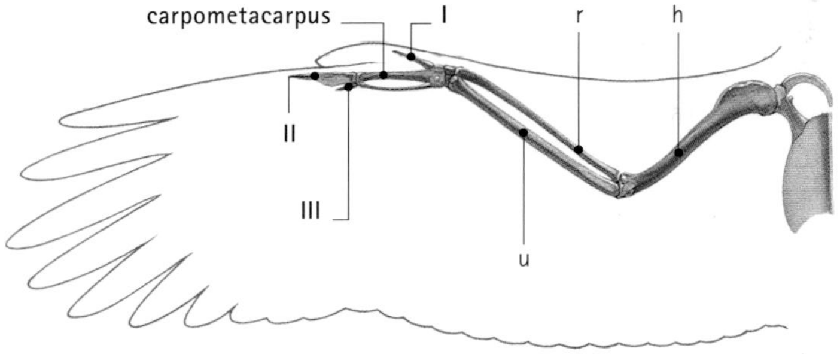

▶ **Bald eagle wing**

The arm bones support the wing surface. The carpal and metacarpal bones are fused to form the carpometacarpus. The third digit supports a flap of feathers called the alula.

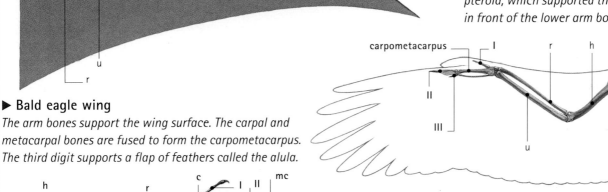

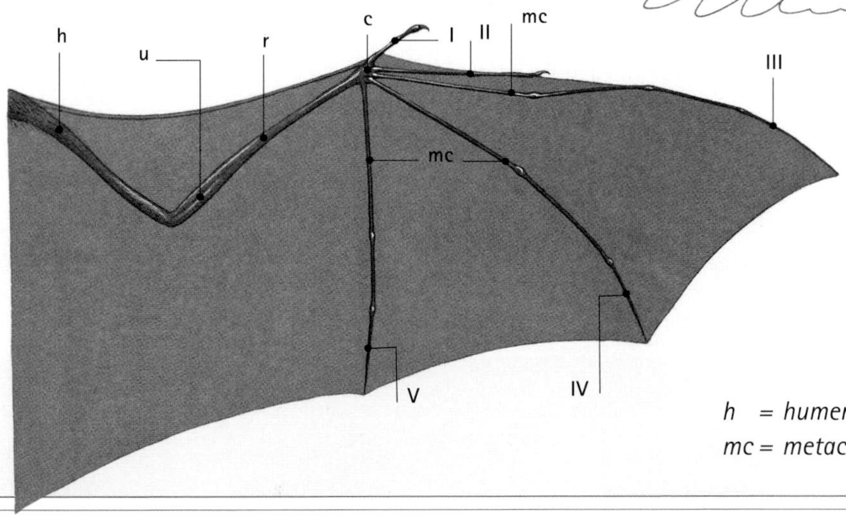

◀ **Fruit bat wing**

Bat wings are supported by the elongated bones of the hand and fingers. The first digit is free from the wing membrane and is used for object manipulation. The second digit has a claw; this feature is unique to Old World fruit bats.

Key

h = humerus *r = radius* *u = ulna*
mc = metacarpal *c = carpal* I–V *= digits*

The spine

A fruit bat's spine is similar to the spine of other mammals, containing 7 cervical vertebrae and 11 thoracic vertebrae. Fruit bats do not have caudal (tail) vertebrae, however, since they have no tail. The thoracic vertebrae are tightly connected and form a stiff column. This helps support the flight muscles.

The pectoral girdle

The pectoral girdle is made up of the scapulae (shoulder blades), clavicles (collarbones), and the sternum (breastbone). Together, they function as the hinge and anchor for the beating wings. Bat wings pivot not at the joint between the arm bones and the shoulder blades, as bird wings do, but deeper in the

IN FOCUS

Locking claws

Fruit bats can hang from a branch by their feet without effort because of an incredible claw-locking mechanism. A tendon runs from the base of each claw to the foot bones in such a way that the pull from the weight of the bat encourages the claws to close up and grip the branch. To make doubly sure, the tendon passes through a tunnel of tissue with knobs that match bumps on the tendon. The knobs pass over each other like a ratchet—easily forward, but with more resistance in a backward direction. Once the claws are closed, they are difficult to open. Bats can sleep and even die while hanging, yet remain attached. A bat must flap its wings to ease the weight on its claws to release them before taking off.

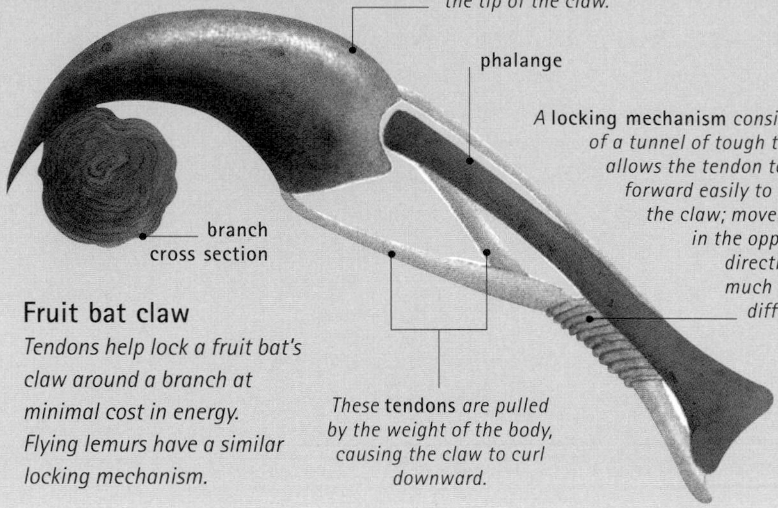

A horny sheath covers the tip of the claw.
phalange
A locking mechanism consisting of a tunnel of tough tissue allows the tendon to slip forward easily to close the claw; movement in the opposite direction is much more difficult.
branch cross section
These tendons are pulled by the weight of the body, causing the claw to curl downward.

Fruit bat claw
Tendons help lock a fruit bat's claw around a branch at minimal cost in energy. Flying lemurs have a similar locking mechanism.

body, at the joint between the collarbones and the breastbone. The shoulder blades themselves are very mobile and move as part of the wing. This gives bats both great agility and stability in flight. Since large bats such as flying foxes do not need great agility, they do not have such specialized pectoral girdles as the more agile microbats.

The pelvis

The pelvis, or hipbones, of mammals usually forms a sturdy frame connecting the hind legs to the spine. The pelvis has two bowl-shaped openings called the acetabula. They act as sockets for the ball-shaped end of the femurs (thighbones), forming the hip joint. In bats the pelvis is even sturdier. The pelvic bones—the ischium, ilium, and pubis—are largely fused. Together with the unusually stiff spine, the pelvis provides a strong anchor for all four limbs and a framework for the muscles to transfer force to the wings.

Leg bones

Bat legs have become adapted through natural selection for hanging instead of support. The legs are good for pulling but not for pushing. The hip joint points directly backward, toward the outstretched feet, not downward as in most mammals. Because of the angle of the hip joint, a bat cannot brings its legs directly beneath the body as other mammals do. When it walks, its legs stick out to either side like those of a reptile. Bat legs are very flexible. The thighbone points forward and sideways during crawling, sideways alone in flight, and backward during hanging. Uniquely among mammals, the knee faces backward; the leg bones have rotated nearly 180° relative to the pelvis. Fruit bats use their legs in clambering more than microbats do. Their legs are relatively larger and more powerful, with thicker bones and more substantial muscles.

▼ Fringe-lipped bats are unusual microbats that often feed on frogs rather than insects. They find prey by listening for the frogs' mating calls. The bat's hand and finger bones are spread wide. This maximizes the surface area of the wing, providing the extra lift required to carry the prey.

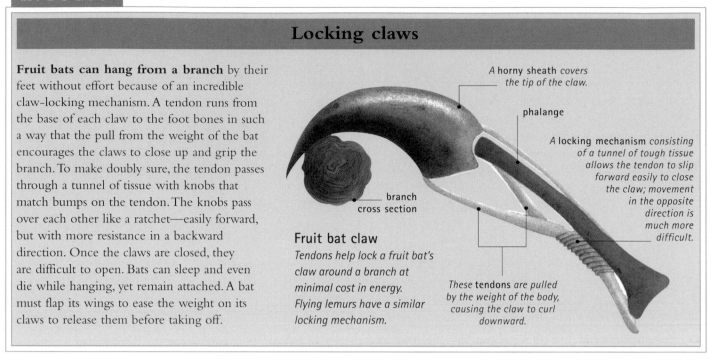

Muscular system

The largest and most powerful muscles in a bat's body are those that control its wing beat. The wing beat is composed of the downstroke, or power stroke; and the upstroke, or recovery stroke. The downstroke takes more energy and demands the larger muscular output. The wing beat is powered by the muscles of the chest and shoulders, muscles that other mammals use to support their body weight on their forelimbs.

If a bat is to be agile in flight, and if it is to flap its wings quickly, most of its muscle mass must be near the animal's center of gravity. This is the point through which the weight of the animal passes. With its muscle-dominated body mass concentrated near its middle, a bat can twist up, down, and sideways, and flap its wings quickly and with minimum effort. A bat's wings and legs are spindly, bearing only wafer-thin strips of tendon and muscle. This helps increase the animal's maneuverability.

IN FOCUS

Bat heart muscle

The energetic demands of flight have led to modifications of the bat's heart. Relative to body mass, a bat's heart is the largest and most muscular of any mammal, with the densest network of coronary capillaries. It weighs between 0.6 and 2.3 percent of total body weight—proportionately two to three times larger than a mouse's heart. Bat cardiac (heart) muscle is constructed from extra-thin cardiac muscle fibers. The fibers are more densely packed and deliver more power than those of other types of mammals. Furthermore, the highest concentration of adenosene triphosphate (ATP; chemical energy) reserves in any mammalian muscle tissue occurs in bat hearts.

Fold up and extend

A bat folds up and extends its wings almost by remote control, actively tensing only the muscles within its trunk and letting a kind of

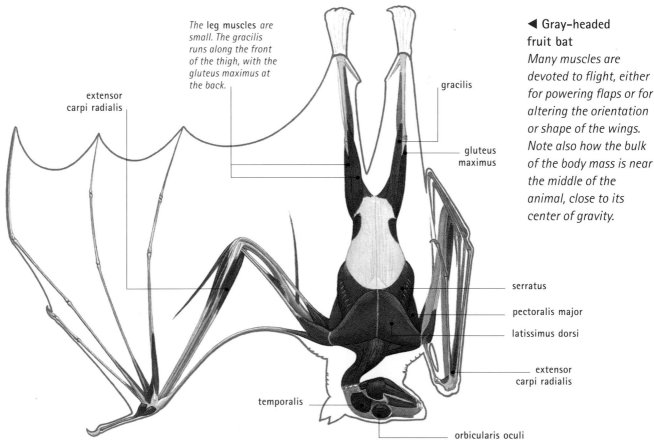

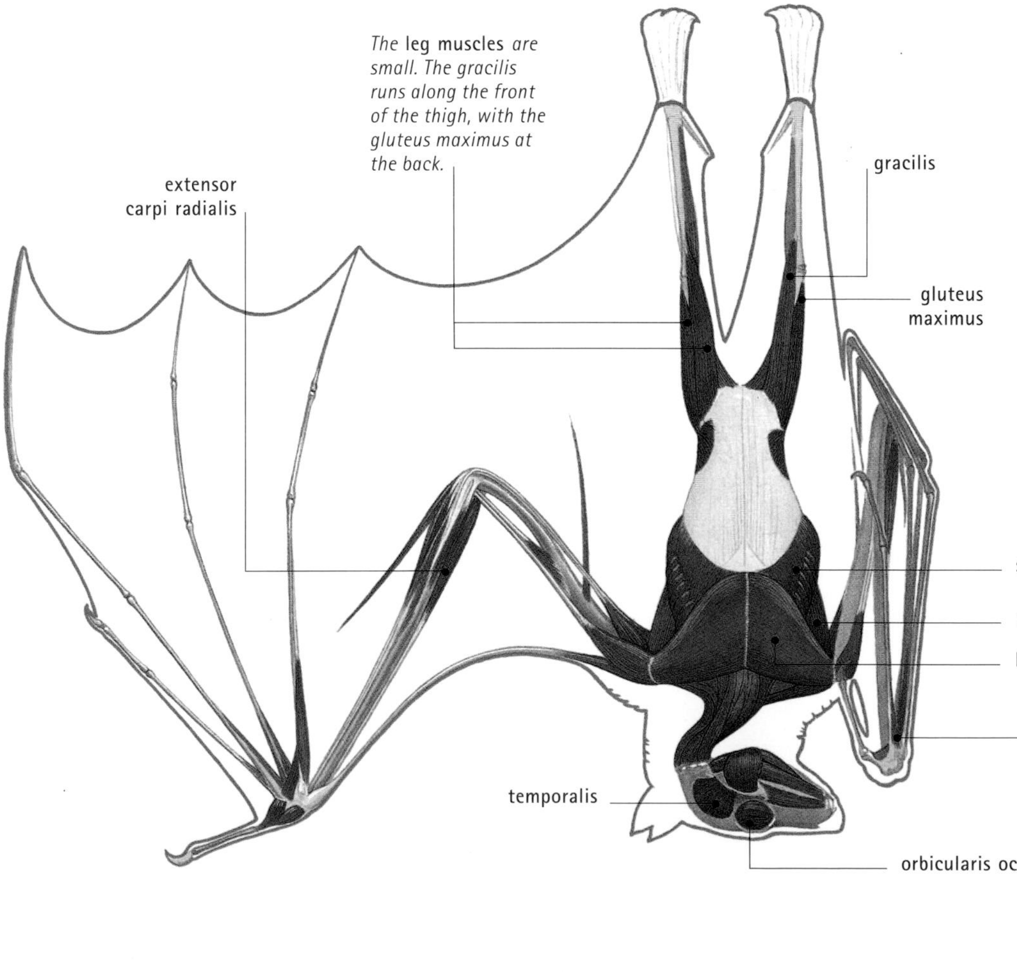

◀ Gray-headed fruit bat

Many muscles are devoted to flight, either for powering flaps or for altering the orientation or shape of the wings. Note also how the bulk of the body mass is near the middle of the animal, close to its center of gravity.

pulley system do the rest. To extend a wing, a bat tenses the supraspinatus muscle, which lies between the scapula (shoulder blade) and the humerus (upper arm). This sets up a chain reaction of passively contracting muscles connected by tendons, which causes the spoke structure of the wing to snap open. A similar chain reaction reverses the procedure. The bat needs only to contract the shoulder muscle (teres major). A cascade of muscular and tendon movements then folds the wing. The cascading movement systems allows the massive trunk muscles to do all the work; muscle mass in the wings is kept to a minimum, so most of the mass can remain near the center of gravity.

IN FOCUS

Leading-edge flaps

The occipito-pollicalis muscle runs from the base of the head to the wing. There, it connects to a tendon that continues along the leading edge of the wing all the way to the second finger, just beyond the thumb joint. At slow flight speeds, the muscle tenses up, and the leading edge of the wing dips, just like the leading-edge flaps of an airplane. This increases the angle of the wing, generating more lift to keep the bat from stalling.

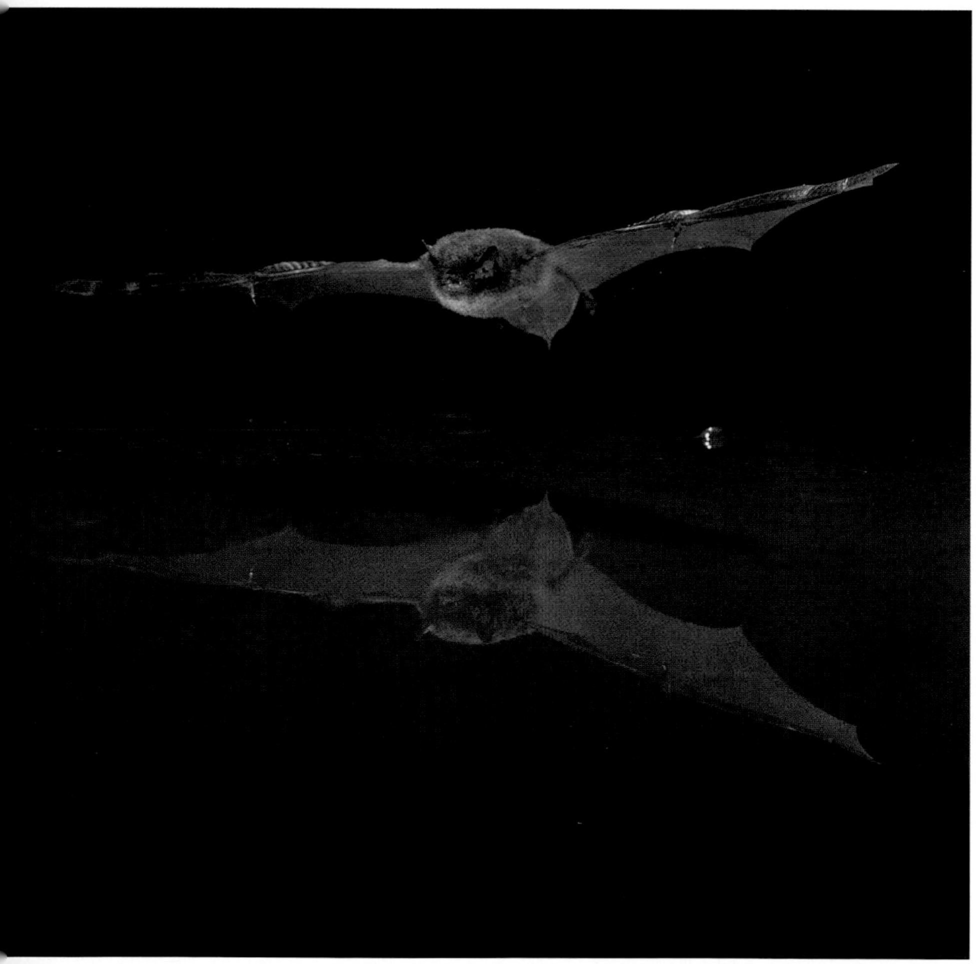

▼ *Like any other animal, a bat tries whenever possible to minimize the energy used by its muscles for movement. Many bats, such as this Daubenton's bat from Europe, often make use of an aerodynamic phenomenon called ground effect. By flying close to a smooth, flat surface, such as this forest pool, the bat can significantly reduce the number of flaps it needs to make.*

Adjusting shape and camber

Bats have unique cutaneous muscles in the wing that adjust the shape of the wing precisely. Some cutaneous muscles keep the wing taut by pulling the wing in toward the trunk, countering the outward pull of the fingers near the wing tips. Other cutaneous wing muscles adjust the curvature, or camber, of the wing. Especially important is the fifth finger, which points back across the wing from the wrist; and its muscle, the adductor digiti quinti. When the muscle contracts, the fifth finger flexes to increase the wing's camber. High camber gives more lift at slow flight speeds but produces a lot of drag. So when the bat is flying fast, the adductor digiti quinti relaxes to give a flatter, low-drag wing shape.

COMPARATIVE ANATOMY

Powerful pecs

Birds have only two massive muscles that control the wing beat—the pectoralis and the supracoracoideus. By contrast, bats use 17 muscles to control their wings. Most of these change the orientation of the wing, altering features such as angle. Only a few powerful muscles control the wing beat itself. In both bats and birds, most of the power to drive the downstroke comes from the pectoralis muscles, which run along the length of the sternum (breastbone) and attach to the upper arm, pulling it down. A bird's pectoralis muscle is enormous; it is attached to the huge plate on the underside of the sternum, the keel. A bat's pectoralis is much smaller, but it still weighs four times more than the other flight muscles. The subscapularis and parts of the deltoid and serratus anterior help power the downstroke.

Nervous system

An animal's nervous system passes messages quickly around the body, allowing the animal to respond rapidly to its surroundings. The sources of these messages are often the sense organs, which capture information from different parts of the animal's body and from the outside world. Sensory nerves transmit messages from the sense organs to the central nervous system (CNS), which consists of the spinal cord and brain. The CNS processes the information and decides how to respond. It sends instructions down motor nerves, which stimulate body parts such as glands and muscles to act on the sensory information.

Fruit bats have amazing sense organs and a remarkable CNS. The bats fly in forests at night, orienting themselves and avoiding obstacles by sight. Many fruit bats have complex social lives, and they have an excellent ability for visually guided reaching and grasping.

Fruit bats' vision

An animal's visual system is often measured in terms of resolution, or sharpness—the sharper the better. A fruit bat's eyes must work best at night, when little light is available, but its eyes cannot maximize sharpness and light-gathering ability at the same time. The eye's pupil (the hole at the front through which light passes) can open wide to let the maximum light in, but this causes the image produced to lose sharpness. The bat's visual system represents a compromise between sharpness and illumination; resolution is sacrificed but lots of the scarce nocturnal light can be gathered. Fruit bats compensate partly by having very large eyes relative to body size, as do other nocturnal animals such as bush babies and owls. Large eyes increase image sharpness because they have more light-sensitive cells on the back of the eye.

CONNECTIONS

COMPARE the eye of a fruit bat with that of a primate such as a ***HUMAN*** or ***MANDRILL***. Similarities with night-adapted eyes suggest that primate ancestors may have been nocturnal.

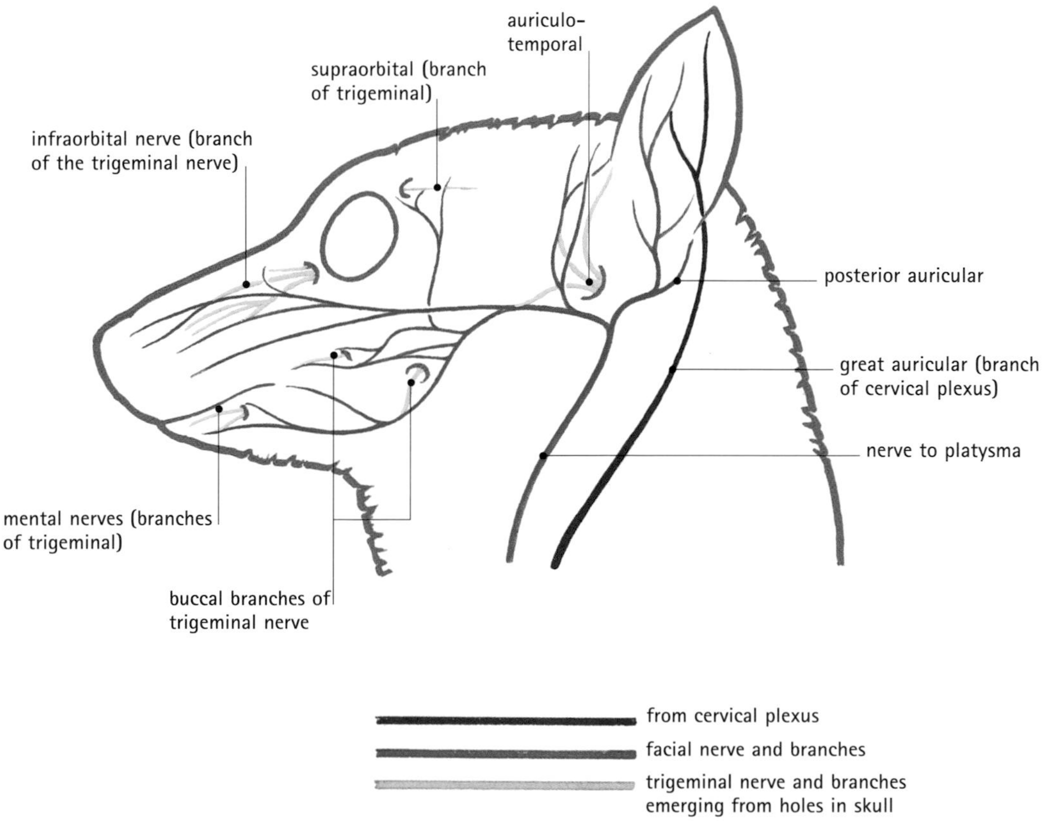

◀ Gray-headed fruit bat

Major nerves inside the head of a fruit bat. The major sense organs—the eyes and sensory epithelia in the nostrils—have particularly numerous nervous connections to the brain.

Fruit bats' eyes have rounded lenses, which are best for focusing on nearby objects. The bats are therefore nearsighted, as are most other nocturnal animals. Fruit bats' lenses also offer high depth of focus, so more distant objects can also be observed in sharp detail.

Sensitive retinas

The retina is the light-sensitive layer at the back of the eyeball. In fruit bats, it is densely packed with "rod" light receptor cells. There are up to 430 million rods per square inch (672,000 per square mm), more than four times the density in a human eye. Rods are extremely sensitive to light intensity, but they cannot detect color—fruit bats are probably color-blind. This is not surprising; color detection is of little use for a nocturnal animal.

Fruit bats are unique among mammals because their retina is supplied with blood by a "choroid" structure, like a placenta. Fingers of the choroid structure intrude upward into the retina. The retina is buckled like a volcanic landscape as a result. This is surprising, since experts thought that a retina needed to be flat for effective focusing. The bumps in a fruit bat's retina could be an adaptation for detecting movement in low-light conditions.

CLOSE-UP

Binocular pathways

Having forward-facing eyes allows fruit bats to have binocular vision. The visual field of each eye (what the eye sees) overlaps greatly with that of the other eye, so a wide arc of the bat's surroundings can be seen with both eyes at once. A bat's brain uses input from both eyes to calculate the distance to objects—in other words, it can see in three dimensions, including the dimension of depth. In this sense, fruit bats are like humans and other primates. Binocular vision is also enhanced by the visual pathways in a fruit bat's brain. The optic nerve from each eye branches into two. One branch travels to the near side of the brain; the other crosses to the opposite side. Each side of the fruit bat's brain receives input from both eyes, aiding in depth calculations. Fruit bats and primates have this crossing arrangement, but echolocating bats do not. Intriguingly, rousette fruit bats, the only Old World fruit bats to possess echolocation, have simple pathways from each eye to each near side of the brain, just like echolocating bats and most other types of mammals.

▼ CROSS SECTION OF EYE
Fruit bat
The round shape of the lens and cornea enables the bat to focus only on nearby objects.

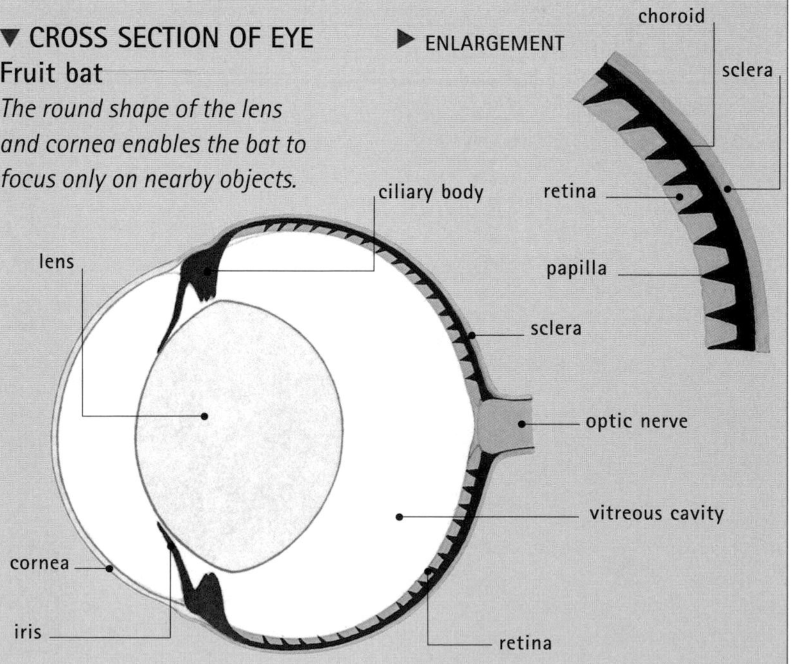

Visual pathways

From the light-sensitive cells in the retina, visual information is transmitted along neurons (nerve cells). Cords of neurons form the optic nerve, which leads to the brain. In fruit bats most visual information travels to the forebrain and then on to the visual cortex. Scientists have traced the path of nerves from different parts of the retina to the visual cortex, and have uncovered how the bat prioritizes information from different sources. In the visual cortex there are far more cells that respond to the middle of the retina, which corresponds to the middle of the visual field. Fruit bats move their head and eyes to keep an object of interest in the center of their visual field, because that is where their processing is most effective.

A separate part of the visual cortex deals with the bottom half of the visual field. This allows a fruit bat to monitor the terrain beneath it when it flies.

The importance of smell

Fruit bats detect their food mainly by smell. Vision guides them only in the final approach to the fruit, so smell is very important. They have a far more sensitive sense of smell than insect-eating bats, especially for acids and alcohols. A flying rousette fruit bat can find 0.1 gram of mashed banana, and is able to distinguish it from banana oil.

The nasal cavity of mammals is lined by a sensitive layer called the olfactory epithelium. It consists of a carpet of tiny projections called

Many rain forest plants are pollinated by fruit bats. Their flowers often open only at night and are usually white; fruit bats do not have color vision, so the plant does not need to waste energy producing pigments in its petals. Fruit bats rely on scent to find food. Bat-pollinated flowers often have a powerful, musky smell to attract them.

microvilli, which increase the surface area enormously. Within the epithelium are a range of different cells, each sensitive to one of a number of different chemicals. Fruit bats' olfactory epithelium is five times thicker than that of insect-eating bats, and much larger in area. It covers complex plates of bone within the nasal cavity, increasing the surface area of the epithelium even further.

At the back of the nasal cavity is a junction with the brain called the cribriform plate, through which the olfactory nerves pass to transmit smell information to the brain. Flying foxes and other fruit bats have cribriform plates that are up to seven times larger than those of insect-eating bats. Powerful as a fruit bat's sense of smell may be, it is still much less effective than that of many other scent-sensitive mammals, such as dogs.

The bat brain

Bats generally have small brains relative to other mammals, but their brains are much larger than those of insectivores such as moles and shrews. Among bats, fruit bats have the largest brains relative to body size. The largest part of a fruit bat's brain is the cortex. The frontal part of the cortex of humans is responsible for judgment, personality, and planning. In fruit bats, the large cortex is probably connected with their complex social lives in their large colonies and with the precise control they need for visually guided grasping and reaching. Similar reasons may underlie the large brain size of primates.

COMPARATIVE ANATOMY

Comparing echolocation

Most bats possess an amazing mechanism for guiding themselves in complete darkness. It is a kind of biological sonar called echolocation. The vast majority of echolocating bats are microbats—small, mainly insect-eating bats quite unlike fruit bats. Microbats send out chirps from their larynx. The chirps are a type of ultrasound, which are of such high frequency that human ears are unable to detect them.

The chirps reflect off objects, and the bat processes the returning echoes to construct three-dimensional sound images of its surroundings. Fruit bats do not generally possess echolocation and rely instead on sensitive night vision. The only exceptions are the rousette fruit bats, such as the Egyptian fruit bat. These bats roost in pitch-dark caves and tombs, where vision is useless. They have evolved an echolocation system independently from microbats. However, the rousette bat system is far cruder. While microbats can hunt minute flying insects by echolocation, rousette fruit bats can locate the walls of their caves by echolocation, but little more. Rousette fruit bats do not chirp to produce echolocation signals. Instead, they produce what sounds like buzzing to the human ear: a rapid train of clicks made by the bat's tongue.

▼ Although it can echolocate, an Egyptian fruit bat relies more on vision, and has very large eyes.

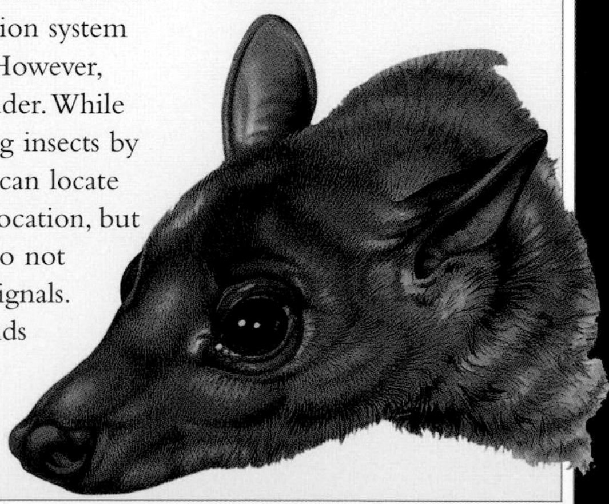

Circulatory and respiratory systems

CONNECTIONS

COMPARE the respiratory system of a fruit bat with that of birds such as an ***EAGLE*** or ***HUMMINGBIRD***.

COMPARE the venous hearts in the wings of a fruit bat with the pulsatile organs at the base of the antennae and wings of a ***HOUSEFLY***.

Respiration is the chemical process by which living organisms release chemical energy from food molecules such as sugars. It generally requires oxygen, and the more energy an animal needs, the more oxygen it requires. Oxygen is pumped into the lungs. It circulates around the body, while a waste product of respiration, carbon dioxide gas, is pumped out. Lungs provide a huge surface over which the bat exchanges gases with the air. An Egyptian fruit bat weighs only 5.6 ounces (160 g), but its lung area is 11 square feet (1 m^2). The diaphragm acts as a bellows; it changes the pressure in the thoracic cavity, causing the lungs to expand and contract.

Lung adaptations

Owing to the high-energy requirements of flight, bats have a great hunger for oxygen. They have a superefficient respiratory system as a result. Bats' alveoli, the minute sacs at the end of the air passages in the lungs, are smaller and

IN FOCUS

A rush of blood?

People once thought that bats needed valves in their arteries to stop blood from rushing to their heads when they hung upside down. In fact, their hanging posture helps blood circulation. Circulation of blood leaving the heart through arteries is not difficult to regulate. The pressure is high and the arteries are muscular to control blood flow. When blood returns to the heart in the veins, however, blood pressure is at its lowest, and circulation could be difficult. A bat's blood is helped back to the heart from the most distant veins, in the legs and wings, by gravity. Blood traveling in veins leading from the head has only a short distance to climb against gravity to reach the heart.

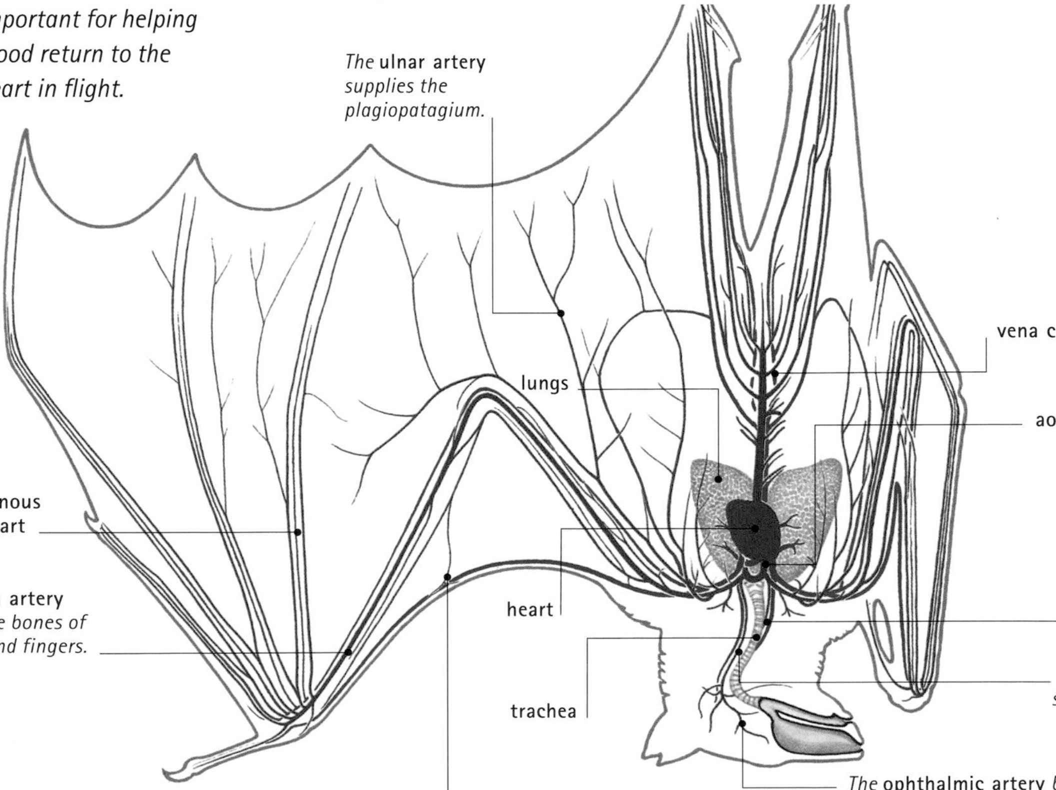

The ulnar artery supplies the plagiopatagium.
lungs
vena cava
aorta
The thyroid artery is a branch of the carotid artery that supplies the thyroid gland. In microbats, this artery is called the laryngeal artery and has the job of supplying the larynx. This is the organ that produces the bat's all-important ultrasonic echolocation pulses.
venous heart
The median artery supplies the bones of the wrist and fingers.
heart
trachea
The carotid arteries supply the head muscles and tissues, as they do in other mammals
The radial artery supplies the arm and propatagium.
The ophthalmic artery branches off the basilar artery to supply blood to the chorions inside the eyes. They then deliver blood to the retinal cells.

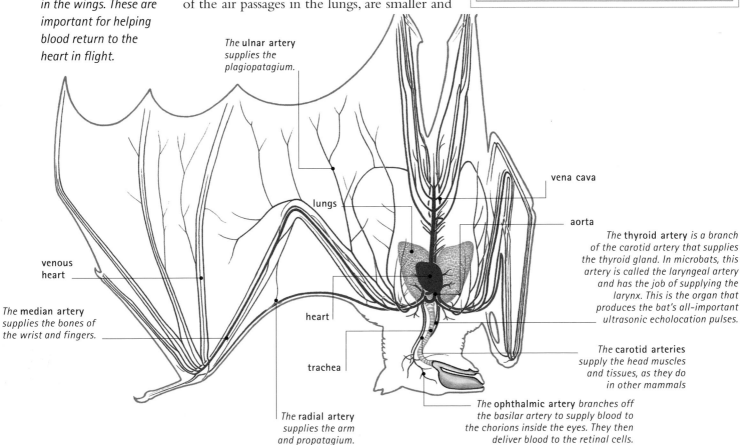

▼ **Gray-headed fruit bat**
A fruit bat's circulatory and respiratory systems. Note the venous hearts in the wings. These are important for helping blood return to the heart in flight.

CLOSE-UP

Venous hearts

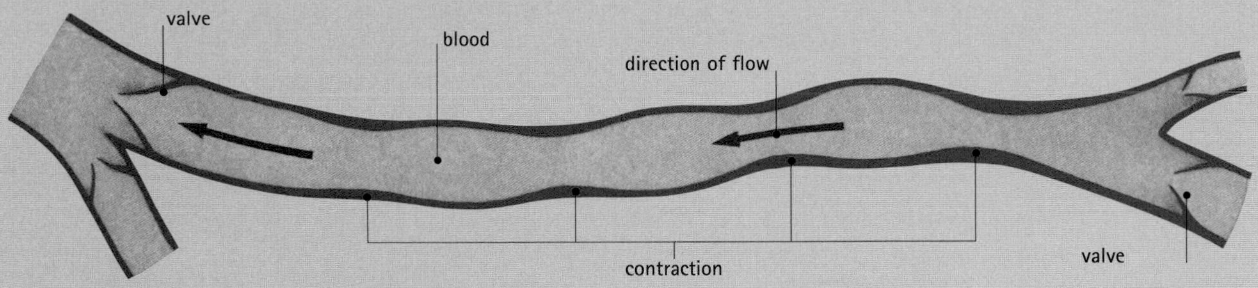

As a bat's wings beat, their rotational movement creates a centrifugal force. This forces blood to the wing tips. To counteract this effect, a bat has rhythmically contracting veins called "venous hearts." These act like little tubular hearts, pumping the blood back to the main heart and overcoming the great distance the blood must return from the wing tips. The veins are lined with normal smooth muscle, like that in other parts of the body but unlike the unique cardiac muscle of the heart. Despite this, the vein muscles contract rhythmically without any input from nerves. Valves prevent blood flowing in a reverse direction and the sections of vein between valves contract independently of each other and can have quite different contraction rates.

more numerous than those of other mammals. They have twice the surface area of a shrew's alveoli and six times that of a chicken's, relative to body weight. The alveoli are also covered with a denser mesh of blood capillaries than those of most mammals. Capillaries are the tiny blood vessels that exchange gases between the blood and the air inside each alveolus. A fruit bat's capacity for diffusion (movement) of gases between its blood and the air in its lungs is higher than that of any other mammal.

Circulating blood

Once bats have obtained oxygen from the air in the lungs, they must distribute it swiftly to the tissues that need it. A bat does this with the circulatory system: the heart and bloodstream. The heart is another pump—not a bellows like the diaphragm, but a hydraulic pump that squeezes blood into the arteries. Arteries are vessels that transport blood to the tissues. They branch into thousands of threadlike capillaries that form a mesh through the tissues, distributing the oxygen in the blood to each cell.

The tissues and organs with the highest metabolism (the fastest chemical processes and therefore the fastest energy consumption) have the densest capillary networks. These organs include the liver, intestines, kidneys, and brain. The organ with the densest network of capillaries is the pectoralis muscle, which needs a constant supply of oxygen and nutrients during flight. It has between 3,600 and 6,400 capillaries per square millimeter—the densest capillary bed in any tissue of any mammal. Blood returns to the heart via a series of vessels called veins.

COMPARATIVE ANATOMY

Gas exchange

Birds do not have as great a lung surface area or capillary density as bats, but birds are better at extracting oxygen from air. This is because they have a one-way flow of air through the lungs. They have a system of air sacs and parabronchi—air passages not possessed by any mammal. This one-way flow makes oxygen extraction much more efficient. In mammal lungs, including those of bats, air passes in and out the same way. Some stale air reenters the lungs with each breath, and some of the air in each lungful does not reach the alveoli; efficient oxygen extraction is not possible. Nonetheless, bats are the only mammals that come close to the oxygen extraction rate of birds.

Digestive and excretory systems

Fruit is a tough diet on which to survive. It is rich in energy, mainly in the form of sugars such as fructose and glucose, but there is little fat or protein. Fruit bats also eat nectar and leaves, but few other mammals eat such a high proportion of fruit. These bats have several unique features that help them cope with food that is so nutritionally unbalanced.

Swift digesters

Fruit bats feed by gripping fruit firmly with their long canine teeth and crushing it against the bony palate (the roof of the mouth) with the cheek teeth and tongue. They swallow the juice and spit out the fibrous pulp. The bat avoids an unnecessary increase in mass (which would cause a serious rise in the energy costs of flight) by spitting out the pulp. This also helps it save time in digestion. A fruit bat's food takes only 12 to 34 minutes to pass through the entire digestive system. Most of the food is in the form of sugars, which are such small molecules that they need very little digestion. The sugars can be absorbed almost immediately into the bloodstream across the walls of the stomach and intestine.

Within 10 minutes, a fruit bat's meal has reached the duodenum (the first part of the small intestine), where much of the digestion and absorption of food takes place. Within another 11

IN FOCUS

Weight gains

By spitting out fruit pulp and swallowing only the juice, a fruit bat takes in around one-third of the fruit's volume and one-third of its energy. However, the bat obtains four-fifths of the digestible energy in the fruit. A fruit-eating monkey also spits out much of its food; only two-fifths of the pulp in its stomach is digestible. Since a fruit bat is so dependent on flight, it needs to keep weight in its stomach to a minimum.

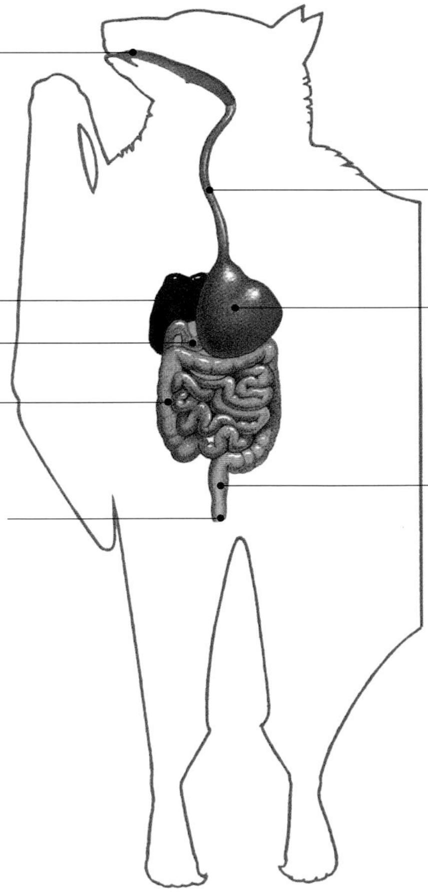

▶ **Gray-headed fruit bat**
In common with other fruit-eating animals, a fruit bat has a relatively short, straight intestine. That is because there is little actual digestion required, since most of their food is in the form of sugars already. Nutrients can simply be absorbed straight across the gut wall.

CONNECTIONS

COMPARE the gut of a fruit bat with that of a ***HUMMINGBIRD***, which also takes a diet rich in easily digested sugars.

COMPARE the cheek teeth of a fruit bat with the carnassials of a carnivore, such as a ***LION*** or ***WEASEL***.

minutes, the waste has passed to the large intestine. There, water is absorbed before the waste is ejected.

A protein deficiency

Although fruit bats digest quickly, it is difficult for them to get all the nutrients they need from fruit. Like all animals, they need to ingest nitrogen-containing chemicals called amino acids, which are the building blocks of proteins. Proteins are a crucial type of biological molecule in any animal's body. Animals usually obtain amino acids by eating protein-rich foods such as meat, nuts, seeds, or leaves.

Both Old and New World fruit bats sometimes eat leaves as a source of protein. They chew the leaves and swallow the juice before spitting out the rest. Fruit contains very little protein, but it does contain amino acids. By picking their fruit carefully, and by eating a lot, some fruit bats are able to meet all their amino-acid requirements through fruit alone.

IN FOCUS

Relaxed kidneys

Kidneys perform the vital task of filtering the toxic waste products of protein breakdown. The waste leaves the body as urine—urea dissolved in water. For most bats, urine should not be too dilute; that would involve unacceptable water loss. That is the other main task of the kidney: to reabsorb water into the bloodstream and to concentrate urine. Fruit bats, however, do not need to concentrate urine. The fruit juice in the gut becomes diluted as sugars and other useful substances are absorbed. The dilute liquid contains plenty of excess water; the bat maintains its water balance not by conserving water, but by urinating freely.

Kidneys are adaptable, though, depending on the conditions. The Egyptian fruit bat normally urinates 0.6 fluid ounce (19 ml) each day. In the desert, however, it reduces this volume 20-fold; the concentration of its urine leaps up by a factor of five.

Many fruit bats also supplement their diets with pollen from flowers, and by eating insects that they encounter crawling on their fruit food. Fruit bats have a relatively low metabolic rate (the speed of chemical processes in the body), so they do not need as much protein as most other mammals.

A diet consisting solely of fruit may also be deficient in certain essential minerals. Some flying foxes may lick salt exuded from glands on the trunks of mangrove trees. Most fruit bats do not have access to mangroves; munching on leaves provides a rich source of minerals such as calcium for these bats.

▲ *An Indian short-nosed fruit bat feasts on a banana.*

▼ SKULL
The fruit bat's cheek teeth help crush up fruit; the tongue and palate (roof of the mouth) combine to squeeze out the juice.

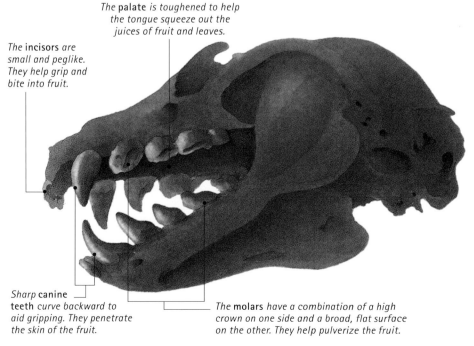

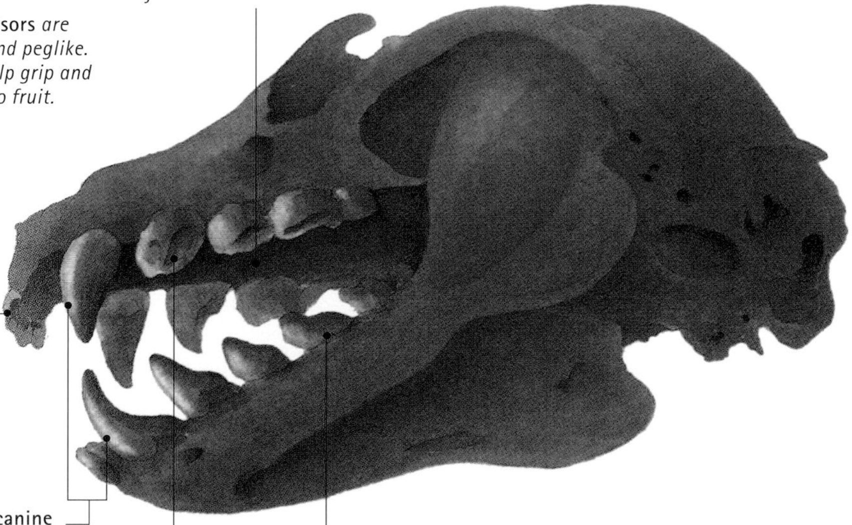

The palate is toughened to help the tongue squeeze out the juices of fruit and leaves.
The incisors are small and peglike. They help grip and bite into fruit.
Sharp canine teeth curve backward to aid gripping. They penetrate the skin of the fruit.
The molars have a combination of a high crown on one side and a broad, flat surface on the other. They help pulverize the fruit.

Reproductive system

The reproductive system of a fruit bat is similar to that of most mammals, but it differs in some details. Fertilization of a female's egg by a male's sperm involves the male inserting his penis into the vaginal canal of a female. The timing of this process must be right. Many fruit bats reproduce only during a breeding season, usually starting in spring. Pregnancy and feeding of young can then occur when food is at its most plentiful.

Hormones control reproductive timing. They are released in response to seasonal changes, such as day length. Hormones cause the testes of male gray-headed flying foxes to start growing in October and the production of sperm to peak in March. This is the time that breeding takes place; if the female is at the right stage of her estrous cycle, an egg is present in the her fallopian tube or uterus. There it meets sperm and may be fertilized.

EVOLUTION

Two uteruses

Experts think that early mammals had a double vaginal canal with a separate uterus, fallopian tube, and ovary at the end of each branch. During mammal evolution, the vaginal canals and uteruses fused to a greater or lesser extent in different branches of the mammal family tree. Modern bats show a range of different conditions, but most fruit bats have either a duplex uterus or a bipartite uterus. A duplex arrangement features two uteruses with separate cervices (the openings at the base of each uterus) that open into a single vagina. A bipartite uterus has two branches but a single cervix. Both arrangements are ideal for mammals with large litters, so it is not clear why fruit bats have duplex and bipartite uteruses. It may, however, reflect fruit bats' ancestry; their nonflying ancestors may have had much larger litters of young.

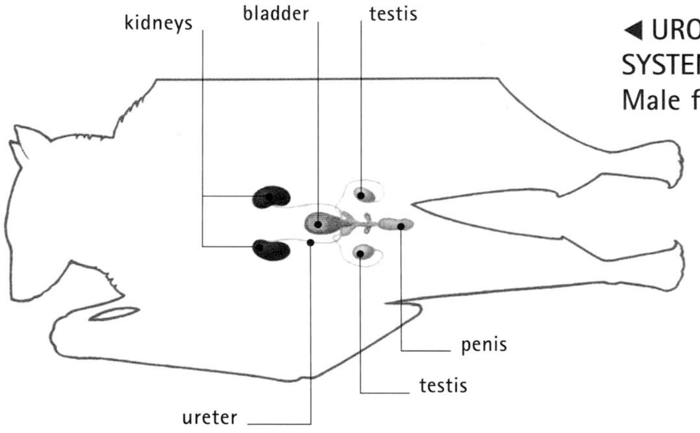

◀ UROGENITAL SYSTEM
Male fruit bat

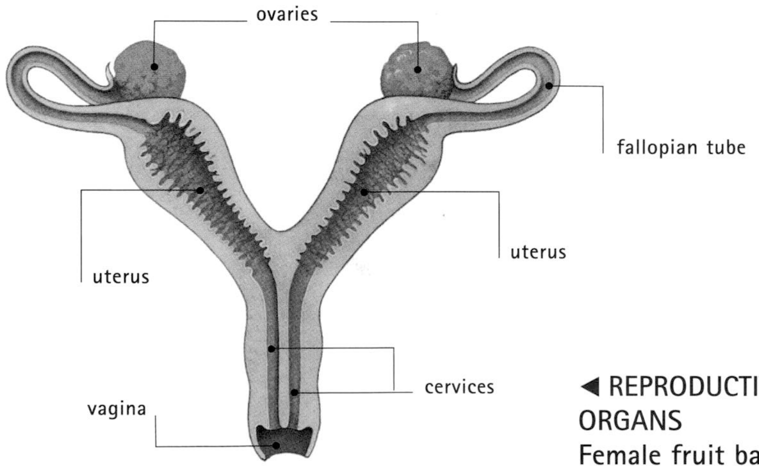

◀ REPRODUCTIVE ORGANS
Female fruit bat

Raising young

The ovaries of most fruit bats alternate in releasing eggs into the uterus. Only one egg is released at any one time, but some African epauletted bats and flying foxes occasionally have twins. Twins result from the release of two eggs. A female gray-headed flying fox carries the developing fetus in her uterus for six months, nourishing it by passing nutrients and oxygen from her blood to its blood via the placenta. At birth, a fruit bat is 12 to 20 percent of the adult's weight and can climb and cling to its mother's fur. For the first three weeks of its life, the pup hangs onto its mother's nipples with its tiny milk teeth as she flies. The mother's nipples are on the chest—another similarity between bats and primates.

The young are not ready to reproduce until they are 1½ to 2 years old. Since only one (occasionally two) young is raised at any one time, the reproductive rate of fruit bats is extremely slow compared with other mammals of their size, such as rats. After one year, assuming no animals die, a pair of fruit bats will increase their number to three or four bats. Over the same period of time a pair of rats would have about 4,000 descendants.

IN FOCUS

Sexual attraction in hammer-headed bats

The fruit bat with the most extreme sexual dimorphism (differences between males and females) is the hammer-headed bat of western Africa. The male is almost twice the weight of the female. He has an enormous square skull, huge pendulous lips, and a warty snout. He has a pair of air sacs in his throat so large that they extend back into the chest cavity. The bony larynx (voice box) is massively enlarged to be nearly half the length of the entire backbone. Together with the enlarged vocal cords and air sacs, the larynx displaces the heart and lungs back and to the side. The male attracts females with a loud honking display. Much of his anatomy is devoted to producing these powerful advertisement calls.

The strange head shape of a male hammer-headed bat helps it produce loud sounds to attract females and intimidate rival males.

Sexual display

The sexes of most fruit bats look similar, but there are many exceptions to this rule. Males are often larger, and they may have larger canine teeth. Some males have conspicuous skin glands, often on their shoulders and chest. The glands produce a smelly secretion that may discolor their fur but plays a role in attracting a mate and defending a territory.

In a few species of fruit bats the males have conspicuous secondary sexual characteristics. For example, male epauletted fruit bats have bright white patches of fur on their shoulders. The patches remain hidden inside pouches of skin for most of the year, but during sexual display an epauletted bat flashes his shoulder patches to attract a female. The males accompany their visual displays with a loud call that they produce with a voice box and throat sacs.

ROBERT HOUSTON

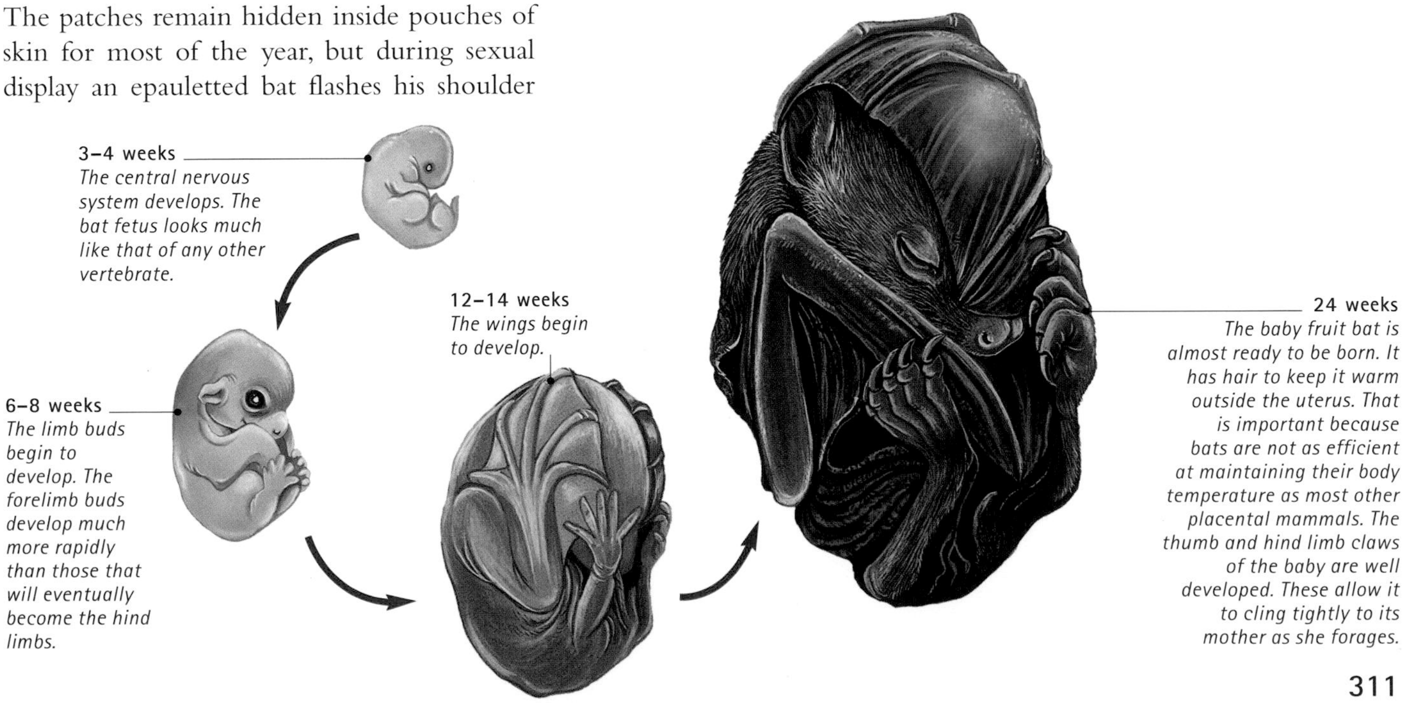

3–4 weeks
The central nervous system develops. The bat fetus looks much like that of any other vertebrate.
6–8 weeks
The limb buds begin to develop. The forelimb buds develop much more rapidly than those that will eventually become the hind limbs.
12–14 weeks
The wings begin to develop.
24 weeks
The baby fruit bat is almost ready to be born. It has hair to keep it warm outside the uterus. That is important because bats are not as efficient at maintaining their body temperature as most other placental mammals. The thumb and hind limb claws of the baby are well developed. These allow it to cling tightly to its mother as she forages.

▼ **EMBRYONIC DEVELOPMENT Gray-headed fruit bat**

Stages of a fruit bat's development in the uterus. The baby bat will be born rear end first. This ensures that the wings do not get caught up in the mother's vagina.

FURTHER READING AND RESEARCH

Kunz, Thomas H. and M. Brock Fenton. 2003. *Bat Ecology*. University of Chicago Press: Chicago, IL.

Macdonald, David. 2006. *The Encyclopedia of Mammals*. Facts On File: New York.

Neuweiler, Gerhard. 2000. *The Biology of Bats*. Oxford University Press: New York.

Giant anteater

ORDER: Xenarthra INFRAORDER: Vermilingua
FAMILY: Myrmecophagidae GENUS: *Myrmecophaga*

All four existing species of anteaters live in South America. The tiny, arboreal (tree-dwelling) silky anteater and the two species of the larger tamanduas live in tropical forests. The giant anteater lives on grasslands. All anteaters have a long snout, a long sticky tongue, and strong claws, which they use to break into the nests of insects such as ants, termites, and bees.

Anatomy and taxonomy

Scientists group all organisms into taxonomic groups based largely on anatomical features. Anteaters are mammals that belong to an order called Xenarthra, along with the sloths and armadillos. Sloths were once placed in the order Edentata, but this name is no longer used.

• **Animals** Animals are multicellular (many-celled) organisms that get the nutrition they need by consuming other organisms. Animals differ from other multicellular life-forms in their ability to move from one place to another, typically by using muscles. Most animals have a nervous system that allows them to react rapidly to touch, light, and other stimuli.

• **Chordates** At some time in its life cycle a chordate has a stiff, dorsal (back) supporting rod called the notochord that runs along most of the length of the body.

• **Vertebrates** The vertebrate notochord develops into a backbone (spine, or vertebral column) made up of individual bones called vertebrae. The vertebrate muscular system that moves the head, trunk, and limbs consists primarily of muscles that are arranged as mirror images on either side of the backbone.

• **Mammals** Mammals are endothermic (warm-blooded) vertebrates with fur or hair made of keratin. Females have mammary glands that produce milk to feed their young. In mammals, the lower jaw is formed by a single bone. Mammalian red blood cells, when mature, lack a nucleus; all other vertebrates have red blood cells that contain nuclei.

▶ This family tree shows the anteaters and their relatives: the armadillos and sloths. There are four species of anteaters, 20 species of armadillos, and five species of sloths.

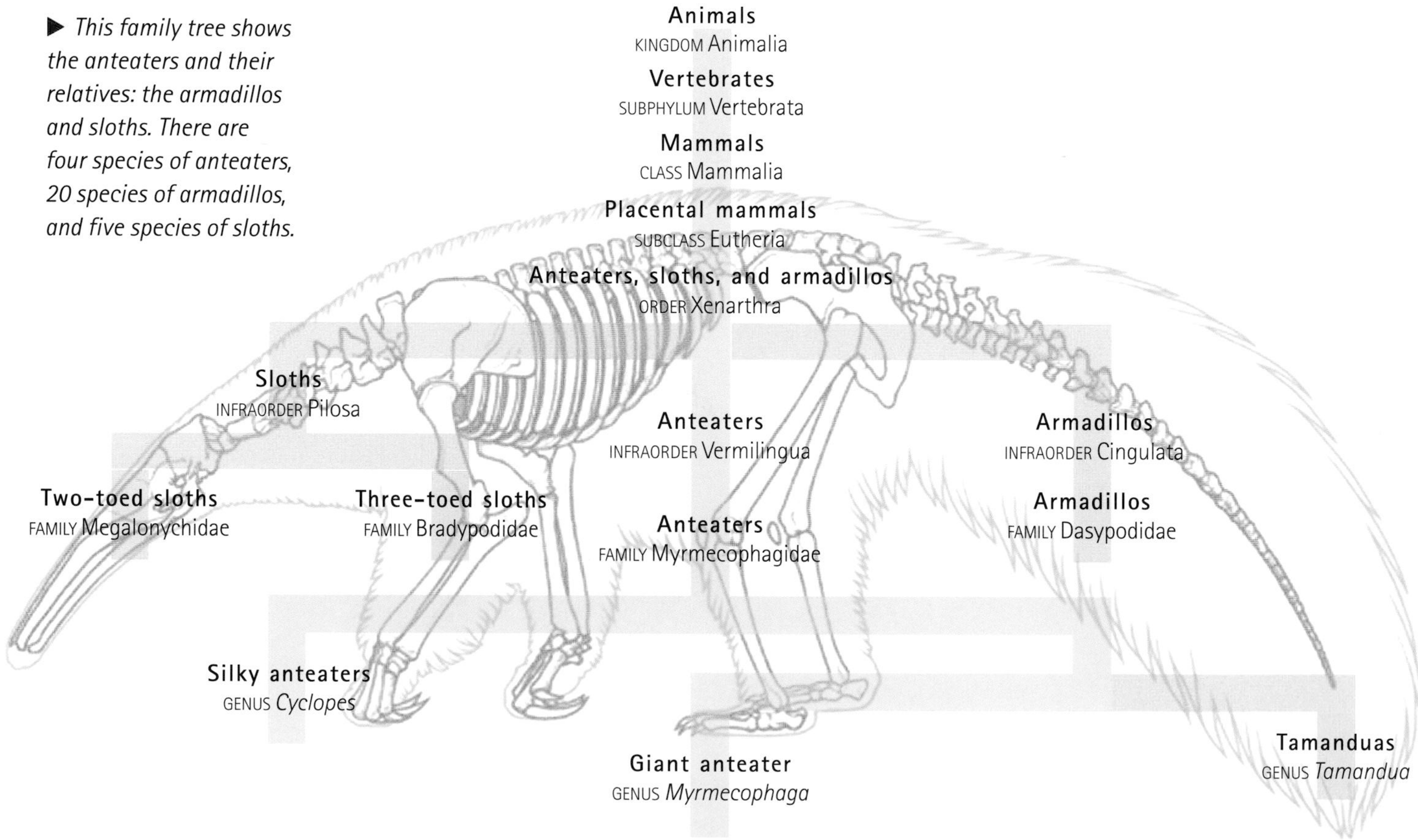

• **Placental mammals** Placental mammals, or eutherians, nourish their unborn young through a placenta, a temporary organ that forms in the mother's uterus during pregnancy.

• **Xenarthra** Xenarthrans are a mainly South American group of mammals, with one species living in North America, where there were many more species in the recent past. All living xenarthrans lack incisors and canine teeth, but only anteaters are completely toothless. In other xenarthrans the remaining teeth are simple and peglike and lack the covering of enamel that occurs on the teeth of other mammals. The number of cervical (neck) vertebrae varies from five to nine depending on the species. This variation is very unusual: mammals, even giraffes, usually have seven neck vertebrae, even giraffes.

• **Armadillos** Armadillos are distinguished from other xenarthrans by their "shell." Covering most of the upper surface of the body, the shell is composed of a series of bony plates covered with a thin layer of horn. Flexible skin connects the plates; this allows some species to roll into a ball when they feel threatened. The limbs and the top of the head are also covered by the horny plates, and the tail is protected by rings of bone. An armadillo's strong pelvis and stout limb bones enable it to dig rapidly, either to make burrows or to find food.

• **Sloths** There are two families of sloths, the three-toed (bradypodid) sloths and the two-toed (megalonychid) sloths. The number of toes refers to the front legs only; the hind limbs always have three toes. Sloths are slow-moving climbers that feed on leaves. These animals spend almost all of their lives in the trees, descending to the ground on a roughly weekly basis to defecate. Members of both sloth groups have long arms, with the claw-bearing forelimbs longer than the hind limbs. Two-toed sloths are covered with pale gray-brown fur that has a greenish hue due to algae that grow on the hairs. Although, like all mammals, sloths are endothermic, their body temperature fluctuates enormously. It can range from as low as 75°F (24°C) to as high as 91°F (33°C). Three-toed sloths even regulate their temperature by basking in the sun, like reptiles.

▲ *The giant anteater is instantly recognizable by its long snout and tongue and the shaggy fur on its body and tail.*

• **Anteaters** As their name suggests, anteaters are specialist feeders on ants, but they also eat other social insects. Anteaters have short, massively powerful forelimbs with long, sharp claws that are used to break into insect nests. Anteaters have a long, tapered snout and a long, sticky tongue to reach inside nests and snare their prey. Anteaters have no teeth. Their large claws make walking tricky; giant anteaters walk on the outside of their wrists; tamanduas spend much of their lives in trees, and silky anteaters are almost exclusively arboreal.

FEATURED SYSTEMS

EXTERNAL ANATOMY Giant anteaters are four-legged clawed mammals with powerful front limbs, an elongated snout, and tiny eyes and ears. *See pages 314–317.*

SKELETAL SYSTEM The skeleton is adapted for breaking into ant and termite hills by using the leverage and physical strength of the forelimbs and claws. *See pages 318–321.*

MUSCULAR SYSTEM The heavily muscled forelimbs contrast with the less well muscled back legs. The long muscular tongue is used to root out social insects from deep inside their nests. *See pages 322–324.*

NERVOUS SYSTEM The giant anteater has a small brain. It is dominated by the region that receives and interprets information from olfactory (smell) sense organs. *See pages 325–326.*

CIRCULATORY AND RESPIRATORY SYSTEMS The chest is small for the animal's size, since the anteater needs to take only small breaths. Blood returns to the heart through two venae cavae. *See page 327.*

DIGESTIVE AND EXCRETORY SYSTEMS Anteaters have no teeth and cannot chew their food. The huge salivary glands are larger than the brain. *See pages 328–329.*

REPRODUCTIVE SYSTEM Female anteaters have a single opening for the urinary and reproductive tracts; the males have internal testes. *See pages 330–331.*

External anatomy

CONNECTIONS

COMPARE the claws of a giant anteater with those of the ***GRIZZLY BEAR*** and ***SLOTH***. The anteater uses its claws to break into ant hills, the grizzly uses its claws to unearth ground-dwelling animals, and the sloth uses its claws to hang from trees.

COMPARE the long tongue of the giant anteater with that of a ***WOODPECKER***. Both use their tongue to pick up insects.

The whole head of a giant anteater is a feeding machine. Ants and termites are easy to find in large numbers, but they are small, and a giant anteater is a large animal. Giant anteaters need to ingest lots of ants and termites very quickly—up to 30,000 ants each day. The giant anteater's head shape reflects its diet and feeding strategy.

Anteaters have an elongated head that extends into a pointed snout called a rostrum. At the end of the giant anteater's rostrum is a small, circular mouth that cannot open very wide. Giant anteaters do not have teeth and can barely move their weak lower jaw. Both upper and lower jaws are vastly elongated. They are held permanently in a tube shape by the skin and muscles of the jaws.

Tongue and snout

The giant anteater's tubelike mouth hides an even longer tongue. An anteater can extend its tongue from the mouth a very long way to

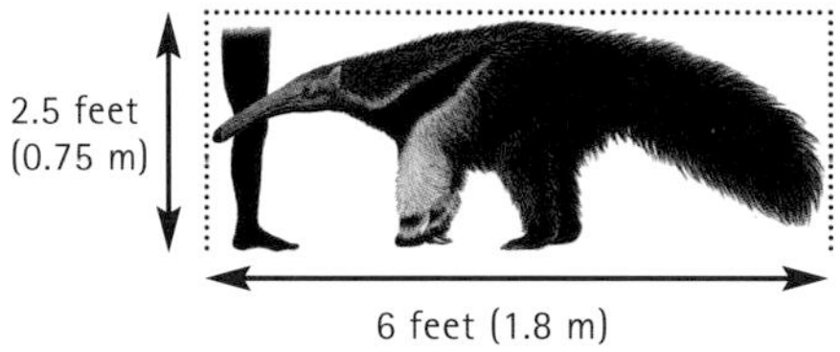

2.5 feet (0.75 m)
6 feet (1.8 m)

The giant anteater bears a distinctive tapering black stripe outlined in white on its flanks.
The snout is long and tubular, and ends in a small mouth opening. The anteater has an excellent sense of smell.
The tongue is around 2 feet (60 cm) long. It is covered by small spines, and large salivary glands keep it coated with sticky saliva. Powerful muscles flick the tongue in and out up to 150 times a minute during feeding.
Vision is less important than smell. Consequently, anteaters have small eyes.
The skin is thick, particularly on the snout, and helps resist the stings of soldier ants and termites.
The massive front claws are flexed and turned inward as the animal walks.
The front claws are used to break through the outer defenses of insect nests.

▼ *Many of a giant anteater's anatomical features are devoted to finding and capturing social insects.*

gather up ants. The snuffling and snorting of a feeding anteater is evidence that suction plays a large part in bringing ants into the animal's mouth. The nostrils are close to the tip of the snout, on the upper side. This brings the scent-detecting cells inside the nasal cavity close to the food, helping the anteater to position its tongue in just the right place. Ants and termites swarm on and around an attacker, so a well-placed tongue inside an anthill gathers up large numbers of insects with every flick.

The coat

An anteater's fur is coarse and thick. It is short on the head and neck, but it becomes progressively longer and looser along the body. The coarse, wavy hairs appear loosely packed, but they form a thick cover and provide an effective shield against rain. Giant anteaters have a distinctive triangle of black fur

▼ Silky anteater
This small, squirrel-size animal lives in dense forests from southern Mexico through the Amazon basin. It spends its entire life among the trees, which it can grasp with its muscular, prehensile tail.

The coarse, stiff bristles of the coat *keep the anteater warm and act as an effective barrier to rain.*

The bushy tail *measures up to 3 feet (90 cm) long. It serves as a counterbalance to the elongated head during locomotion, and as a parasol; it is sometimes held over the body to protect against the sun.*

The hind feet *are placed flat on the ground as the animal walks. This is called a plantigrade stance, similar to that of bears and humans.*

EVOLUTION

Converging on ants

Anteaters appeared around 25 million years ago in South America. Their fossils are almost exclusively confined to the Americas, with one intriguing exception: *Eurotamandua*, a species that lived around this time in Europe. The affinities of this animal are shrouded in mystery.

Pangolins are ant-eating mammals from Africa and Asia. They look like tamanduas with armadillo-like scales and were long considered to be close relatives of xenarthrans; the two taxa were placed together in a group called the edentates, a name which is still seen in some textbooks. However, DNA evidence has shown that pangolins are not closely related to anteaters at all. This is an example of convergent evolution, in which animals evolve similar body plans in response to similar environmental challenges. Similarly, biologists now think that *Eurotamandua* was not actually a xenarthran; however, it did eat ants and evolved in a similar way to the true anteaters.

COMPARATIVE ANATOMY

Claws and feet

A hooklike hand of claws is a feature shared by all anteaters, sloths, and armadillos. Armadillos have claws on the upper side of the foot. They use their claws for digging. Sloths use their claws to hang from branches without expending energy, and to draw food to the mouth. On the ground, a sloth walks on the backs of its hands.

The giant anteater has three large claws and one small claw on each forefoot. The longest claw is 4 inches (10 cm) long; the second claw is half that length. A walking giant anteater tucks its front claws backward against the palms and walks on the outside of its wrists. It uses its foreclaws only when feeding or in defense. Each hind foot carries five smaller claws. A giant anteater walks on flat back feet.

Like the giant anteater, the two species of tamanduas have three long claws on each front foot, used to tear into ant or termite nests. Tamandua claws are not as long as those of giant anteaters, and tamanduas eat insects from smaller nests. Their claws enable them to perform a wider range of tasks; their more varied diet also includes young bees and honey.

Three of the digits on the front feet of a silky anteater are very small. Only one digit carries a large claw. The long claw fits into a groove in a fleshy pad on the opposite side of the palm. This arrangement enables the silky anteater to form a circle like that between the thumb and forefinger of a human. A circular grip is ideal for grabbing onto tree branches while climbing.

The first, fourth, and fifth digits of the front feet are much reduced. The long third digit slots into a fleshy pad, providing an opposable digit that aids grasping.

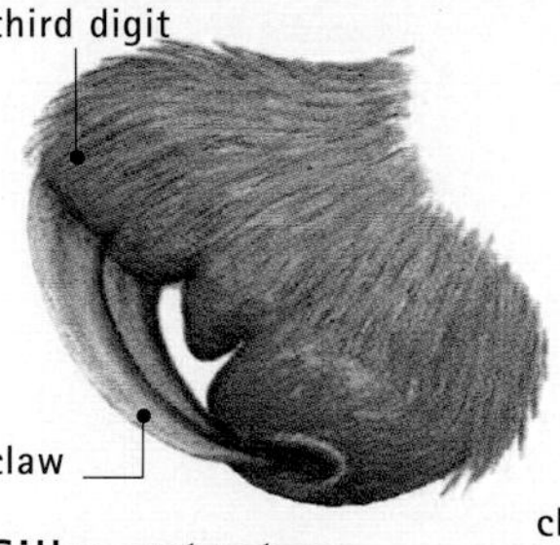

Silky anteater

In both species of tamanduas the middle three digits of the front feet bear sharp claws for tearing into insect nests. The other digits are reduced.

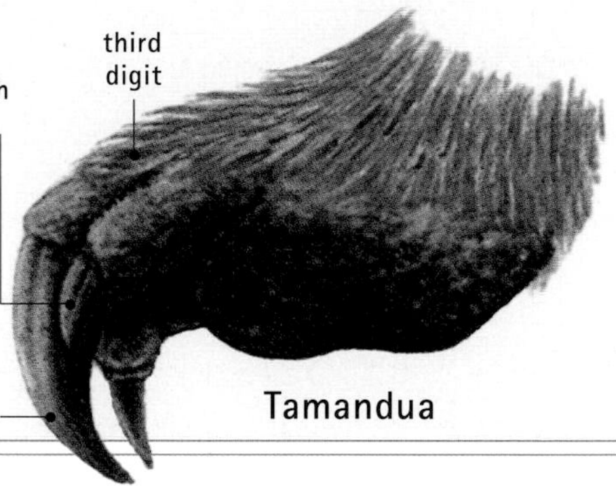

Tamandua

The second and third digits on each front foot have very large claws. When walking, the anteater tucks these claws in, and walks on its knuckles and the sides of the feet.

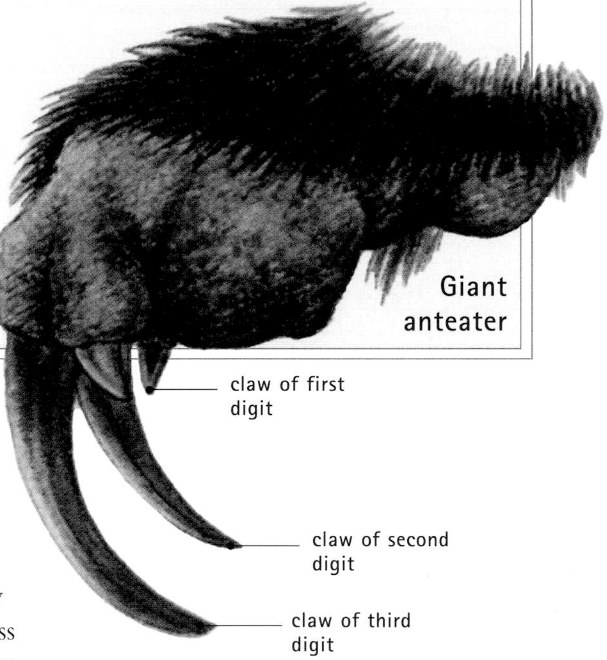

Giant anteater

extending from the lower jaw across the shoulders. Silky anteaters have golden fur. Northern tamanduas are yellow with brown stripes across the chest and on the rump. Most southern tamanduas are a uniform brown, but their color varies: some subspecies are gold; others appear black; and some have a ring of dark fur around the neck. This last species is sometimes called the collared tamandua as a result.

Anteater tails

The tail of a giant anteater supports a long curtain of coarse hairs that are longest nearest the body. When at rest, the giant anteater holds

CLOSE-UP

Antproofing

The head of a giant anteater, including the curved tube of the snout, is protected from ant attacks by a short but dense coat of coarse hairs. The hairs lie flat to the skin and form an effective defense against bites and squirts of acid from angry ants. The short-haired, smooth surface of the long face enables the anteater to wipe its snout free of ants using its curved claws, allowing it to feed a while longer. Anteaters are not immune to ant bites or to ants' chemical defenses. After a time feeding, an anteater needs to retreat from the attacking insects.

◀ **Northern tamandua**
Tamanduas have smoother fur than other anteaters, and their prehensile tail is smooth and hairless. The northern tamandua lives from Mexico through Colombia.

its tail over itself; a sleeping giant anteater keeps its head tucked under its tail for warmth. The tail hair makes an effective parasol in sunshine and an umbrella in the rain. Covered by its bushy tail, the anteater appears as little more than a brown bump in the landscape and can often be difficult to detect in the long grass of the pampas.

A giant anteater can stand on its hind legs while pulling apart an anthill or termite mound with its front claws, or when defending itself against predators. When an anteater adopts this position, it uses its tail as a third standing "limb." The tail steadies the anteater, allowing it to lean backward without tumbling over. Female giant anteaters give birth while standing. The tail provides essential support for the animal at this critical time.

The tamanduas and the silky anteater have a prehensile (grasping) tail that helps them cling to trees. Therefore, in common with the New World monkeys, they have five grasping appendages. The silky anteater wraps its tail around branches and can stand on its hind feet and tail.

▲ **Southern tamandua**
This species of anteater is very similar to its northern cousin, but its coat is more variable in color. It lives in tropical forests of Colombia through Bolivia and Brazil.

PREDATOR AND PREY

Defensive features

Standing tall against an anthill or termite mound, snuffling and ripping chunks from the cementlike nest, a giant anteater is no soft target. The long claws supported by the heavily muscled forearms are formidable weapons. Standing on its back legs and leaning back against the tail, a threatened anteater defends itself by slashing out with its claws or by grabbing an intruder in a hug and slicing the intruder's back with its claws. This dramatic defense strategy helped earn the anteater the nickname "ant bear." Modern adult giant anteaters have no natural predators; however, their defensive strategy may have evolved in response to long-extinct enemies such as saber-toothed cats or short-faced bears. Giant anteaters are not aggressive unless approached closely and threatened. They usually run from danger if they can.

Skeletal system

CONNECTIONS

COMPARE the giant anteater's long, narrow, unopenable jaws with the huge jaws and wide gape of a ***HIPPOPOTAMUS.***

COMPARE the short, strong forelimbs of an anteater with the long, slender limbs of an animal built for speed such as a ***RED DEER.***

The giant anteater's elongated skull is distinguished from that of most other mammals by its simplicity. The jaws do not contain teeth, and the lower jaw is separated at the chin. The gap at the front allows more space for the tongue to flick in and out. The long snout and weak lower jaw are fused into a tube by muscles. The animal cannot open its mouth except at the tip to let out its tongue.

Exploring the skull

The snout, or rostrum, contains the air passages that connect the nostrils at the snout to the large olfactory (smelling) organs at the front of the braincase. The braincase itself is small for a mammal of this size.

IN FOCUS

A tail of two halves

The first few vertebrae behind the pelvis (the sacral and caudal vertebrae) support the thick end of the tail. They are very robust, with long projections that provide attachments for the muscles that allow the animal to use its tail as a support when standing up. Farther along the tail, the vertebrae are progressively thinner and shorter until they reach a fine point. The thin end of the tail needs only to support a curtain of hair.

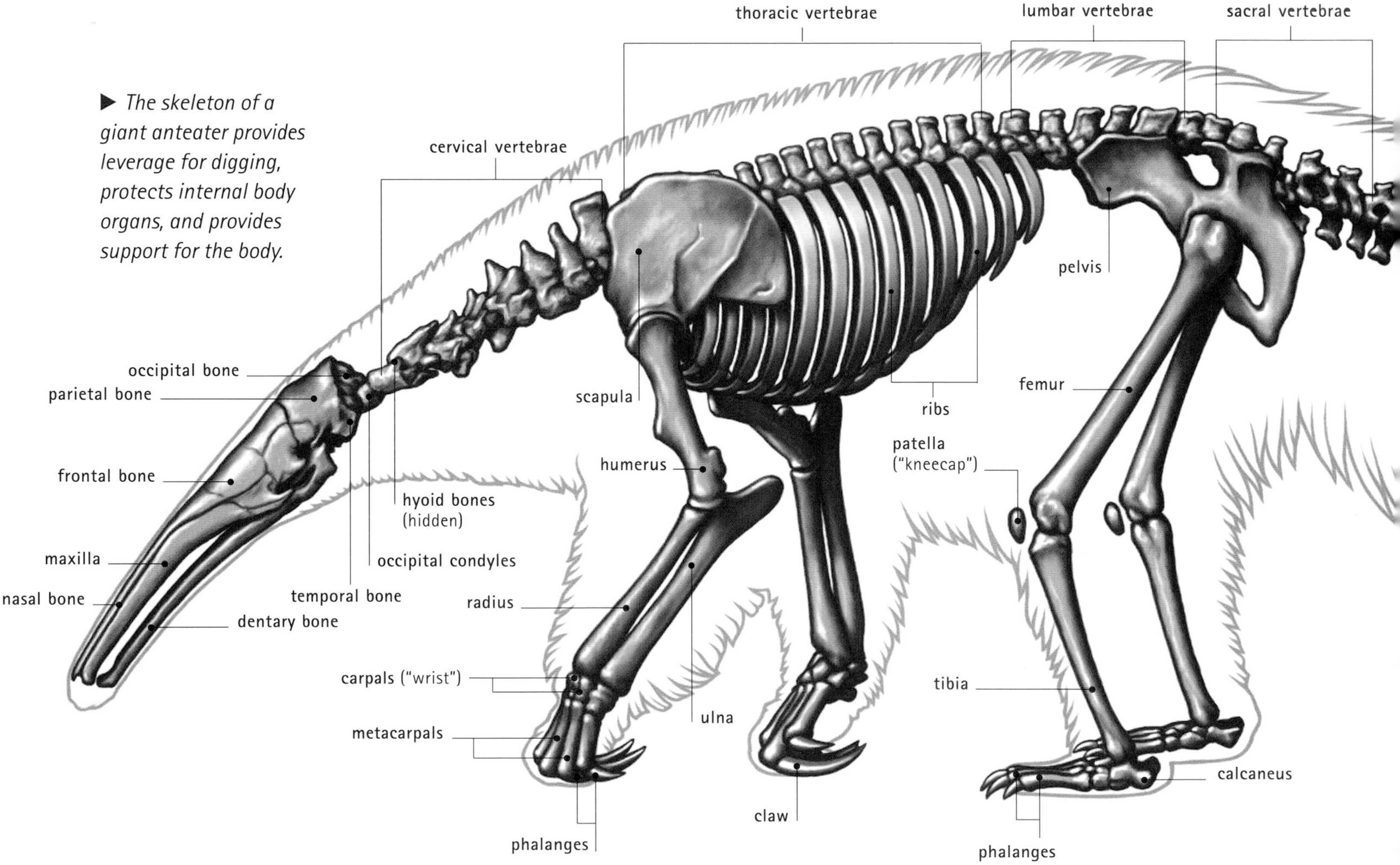

▶ *The skeleton of a giant anteater provides leverage for digging, protects internal body organs, and provides support for the body.*

SKULLS

▶ Giant anteater

The giant anteater has the longest skull of any xenarthran. The jaws are almost fixed in place, only permitting the mouth to open slightly for the projection of the long, sticky tongue.

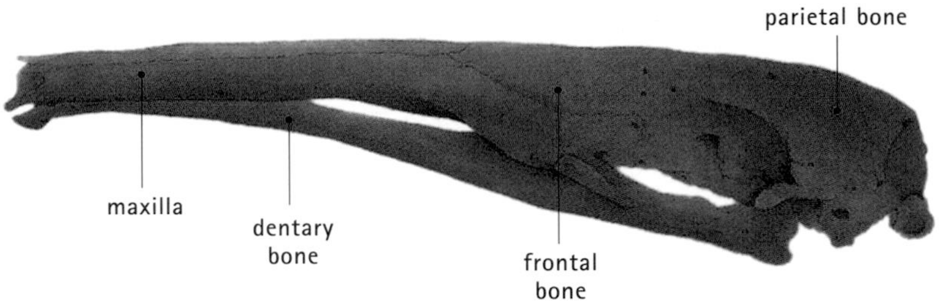

▶ Tamandua

The tamandua's skull has a much shorter snout than a giant anteater's, and the opening for the mouth is very small, only the diameter of a pencil.

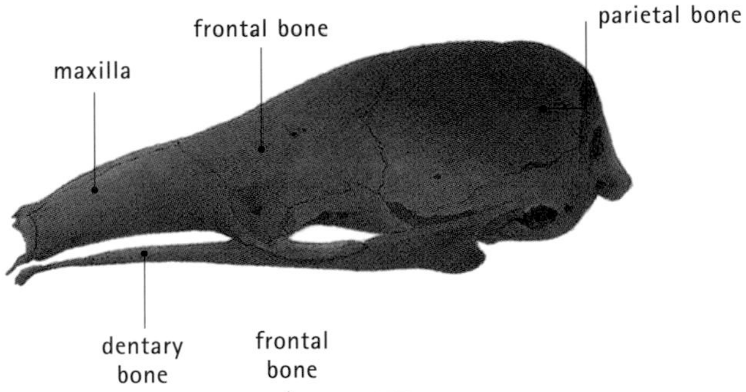

▶ Silky anteater

The silky anteater's skull is much shorter and curves downward. The mouth opening is much larger than that of other anteaters.

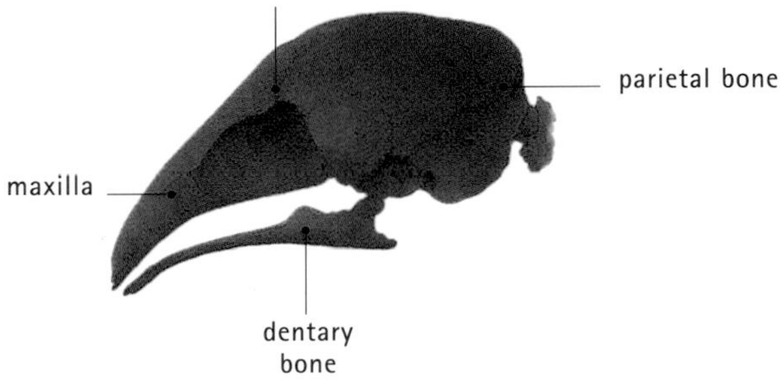

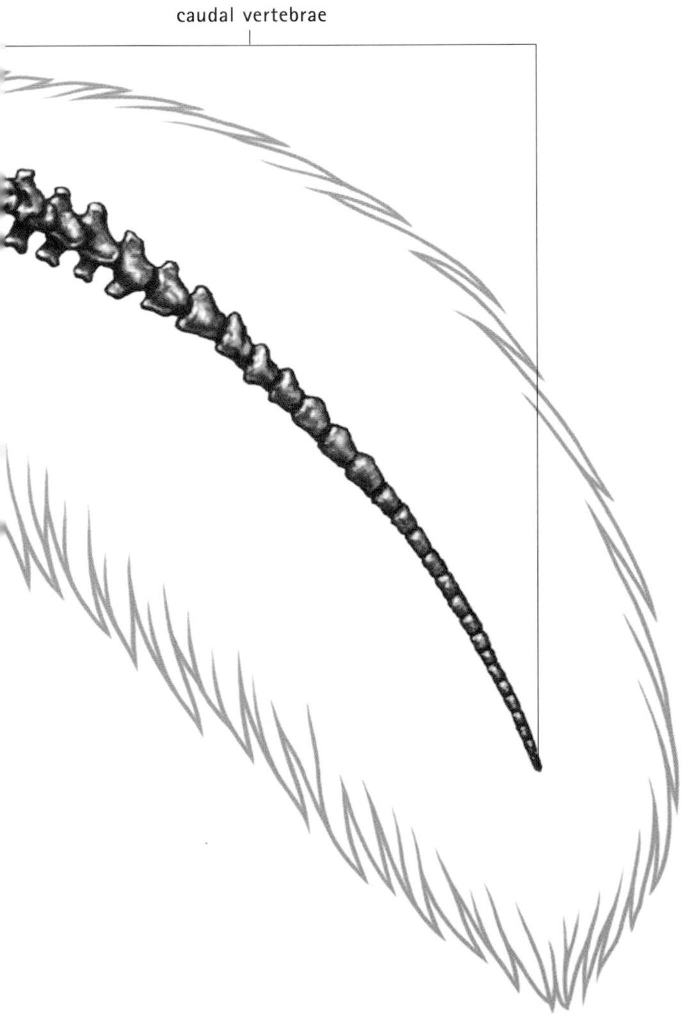

The lower part of the giant anteater's skull includes an extended hard palate. In a human, the hard palate forms the hard roof of the mouth. At the top of the throat, a person's hard palate merges into the soft palate. The giant anteater has a long hard palate that projects backward over the soft palate. Fused bones called the pterygoids support the long, bony hard palate. The hard palate forms the upper side of the tube through which a giant anteater sucks up its food. Muscles supported by the thin mandible bones form the underside of the feeding tube.

CLOSE-UP

Bones that support the tongue

The tongue of most mammals is attached to the throat around the region of the hyoid bones. The giant anteater's hyoid bones are large, supported by muscles, and connected by flexible joints. The supple structure of the bones powers the tongue as it thrusts in and out of the mouth.

IN FOCUS

Anteater vertebrae

The vertebrae of many mammals enable flexibility of the spine. A flexible spine is ideal for speedy running. In a giant anteater, however, the lower back must be held rigid during digging. The vertebrae of the lower back are held firmly by muscles that hold the vertebrae rigid. The rigid back supports the power of the forearms as the animal stands and rips into ant nests. Another important function of the vertebrae is to shield the delicate nerves inside the spinal column. The spinal cord passes through the hole in the center of each vertebra.

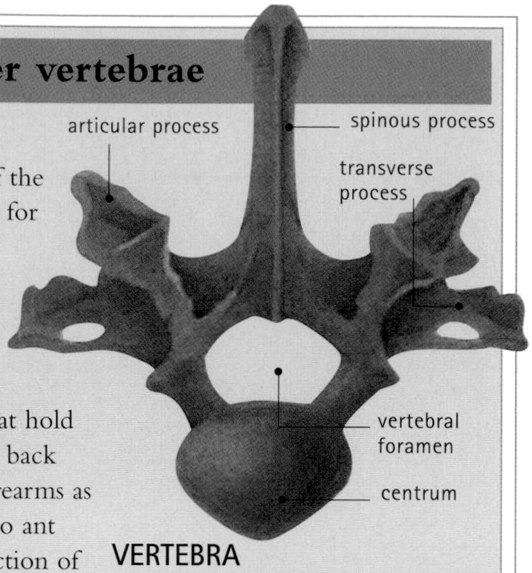

VERTEBRA

The vertebral processes provide attachments for the strong muscles that hold the anteater's back rigid during digging.

The spine

The giant anteater and its relatives have very thick neck vertebrae that support the strong muscles of the neck and shoulders. These muscles both support the head and generate the powerful movements of the animal's shoulders and forearms used to break into insect nests. On the back, between the forearms and the pelvis, the vertebrae are less chunky. The back vertebrae have extra jointlike projections (articular processes) on the top. These extensions of the vertebrae are unique to anteaters. Together, the giant anteater's back vertebrae and the attached muscles make the back unusually rigid. A stiff backbone enables the anteater to exert maximum pressure when tearing down the walls of anthills from an upright position.

Digging adaptations

Like many animals that dig using their forearms, anteaters and other ant-eating animals, such as the aardvark and pangolins,

EVOLUTION

Giant relatives

With the exception of the giant anteater, living xenarthrans are generally small and inconspicuous, being arboreal, nocturnal, or both. However, in the past, their ancestors included some spectacular beasts. Scientists know about them from the fossilized remains of their skeletons. Perhaps the best known were the ground sloths, which lived throughout the Americas. Species such as Jefferson's ground sloth were roughly the size of cows; ground sloths on islands of the Caribbean were the size of a small human. Some ground sloths, however, such as *Megatherium*, were truly massive, up to twice the size of a modern elephant. They browsed on leaves and may have used their large, 7-inch (18-cm) claws to strip bark. All ground sloths had long, sharp claws. They could extend and retract their forelimbs very quickly, making their claws deadly weapons against predators.

Another strange group of edentate giants were the glyptodonts. These car-sized herbivores were among the most heavily armored animals of all time; they were almost completely encased by a carapace of thick bone, with a bony helmet on their head. They also had a bony club for a tail that seems to have been used as a weapon in battles with other glyptodonts. Glyptodonts and the ground sloths of continental North and South America became extinct around 9,000 years ago; some Caribbean ground sloths may have disappeared only around 500 years ago.

▼ Megatherium

The extinct Megatherium *would have weighed around 4.4 tons (4 metric tons). Fossil evidence shows that these creatures sometimes walked upright on their hind legs, placing an enormous strain on their skeleton.*

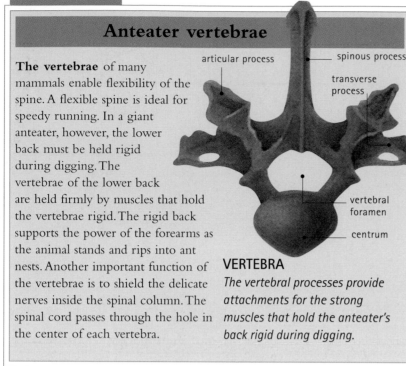

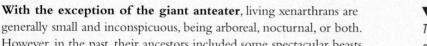

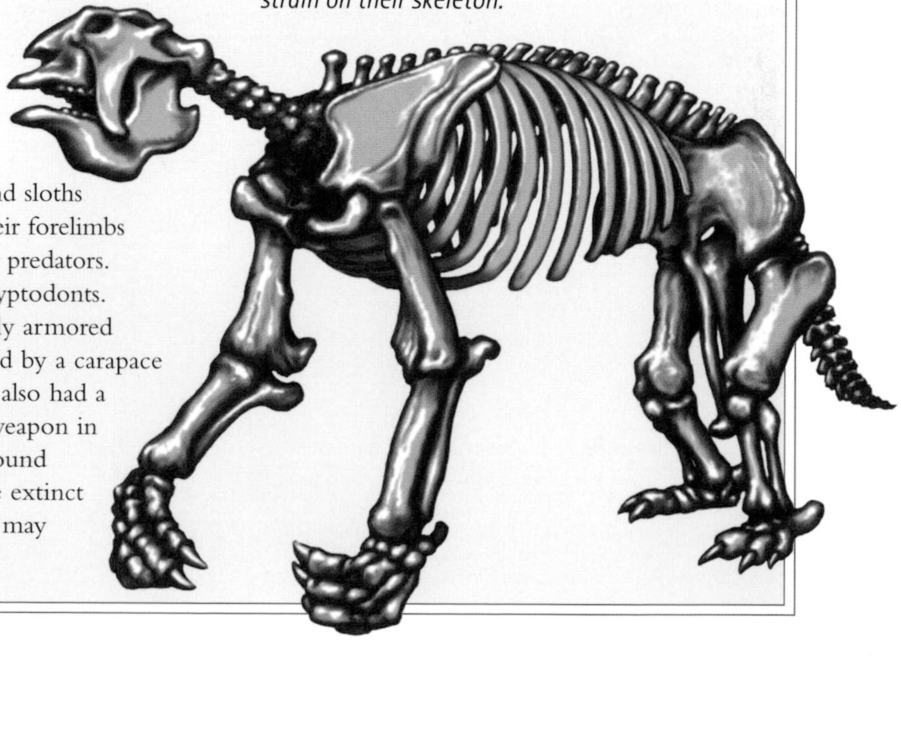

▶ *The giant anteater uses its massive forearms and claws to tear apart the tough soil structures of ant and termite nests. Giant anteaters live in tropical forest, open woodland, and dry savanna in South America east of the Andes mountains.*

have a small pelvis. The pelvic bones and the adjacent vertebrae are fused into a single structure, further adding to the rigidity of the animal's back. The hind legs attach to the pelvis with a ball-and-socket hip joint. This enables the animal to walk on all four legs or swivel to stand on its hind legs, a pose it adopts when threatened or when giving birth.

A giant anteater's forelegs are large and powerful, and specialized for demolition work. Heavily muscled forearms, able to break into rock-hard anthills, need the support of a sturdy set of bones. The fan-shaped scapulae, or shoulder blades, on each side of the upper back are large. The upper arm is short and stout. The shoulder and upper arm provide the strength; the lower arm gives the leverage. The lower arm has long ulna and radius bones. The long phalanges (finger bones) of the hand provide leverage for the anteater's large claws..

COMPARATIVE ANATOMY

Teeth or toothless?

Mammals that eat ants often do not chew and may have only vestigial teeth too small to use, or no teeth at all. Marsupial ant eaters such as numbats have small teeth. Aardvarks and armadillos have simple teeth with no enamel. The most specialized non-chewers are the giant anteaters, pangolins, and spiny echidna, which have no teeth at all. Incisors and canines are used by most mammals to cut and rip food into pieces, but these teeth are lacking in xenarthrans and pangolins, which have just a few molars and premolars. Xenarthrans mostly do not break up their food, so they have no need for incisors or canines. The giant anteater's closest relatives, the tamanduas and the silky anteater, have small, simple teeth with no enamel, and each tooth has only a single root. Tamanduas' and silky anteaters' teeth continue growing throughout their lives. The xenarthran with the most teeth is the giant armadillo, which can have more than 100.

Muscular system

> **CONNECTIONS**
>
> COMPARE the giant anteater's tongue muscles and hyoid structure with those of a *WOODPECKER*. The giant anteater's tongue muscles attach to the sternum, while those of a woodpecker attach to the base of the upper mandible.

Giant anteaters have a thick, heavy neck supported by the vertebrae. Large muscles are attached to the vertebrae, and the muscles support and maneuver the unwieldy, elongated head. Giant anteaters lack the intricate facial musculature required to assume different expressions, but there are many important muscles beneath the surface. For example, the throat contains the hyoid bones and the muscles that power the tongue, in addition to the tongue itself.

A long tongue

The giant anteater has a very long tongue. It is vital for gathering large numbers of adult insects, eggs, and grubs in a short time. Many other animals have a long tongue that aids in collecting insect food. Pangolins and some echidnas, for example, have an erectile tongue containing vascular spaces, which are gaps that can fill with body fluids. The vascular spaces engorge with fluid, stiffening the tongue and making it easier to direct.

Giant anteaters do not have an erectile tongue; they use muscles to point it in the right direction. The giant anteater's long tongue needs to pass through the neck and out into the mouth when in use but must slot fully inside the mouth and neck when at rest. The tongue connects to the top of the chest at two points at the back (posterior) of the sternum, the "breastbone," to which the ribs attach.

▶ SUPERFICIAL MUSCLES

The strong biceps, triceps, carpal flexor, and digital flexor muscles of the giant anteater's forelegs enable it to dig quickly for food.

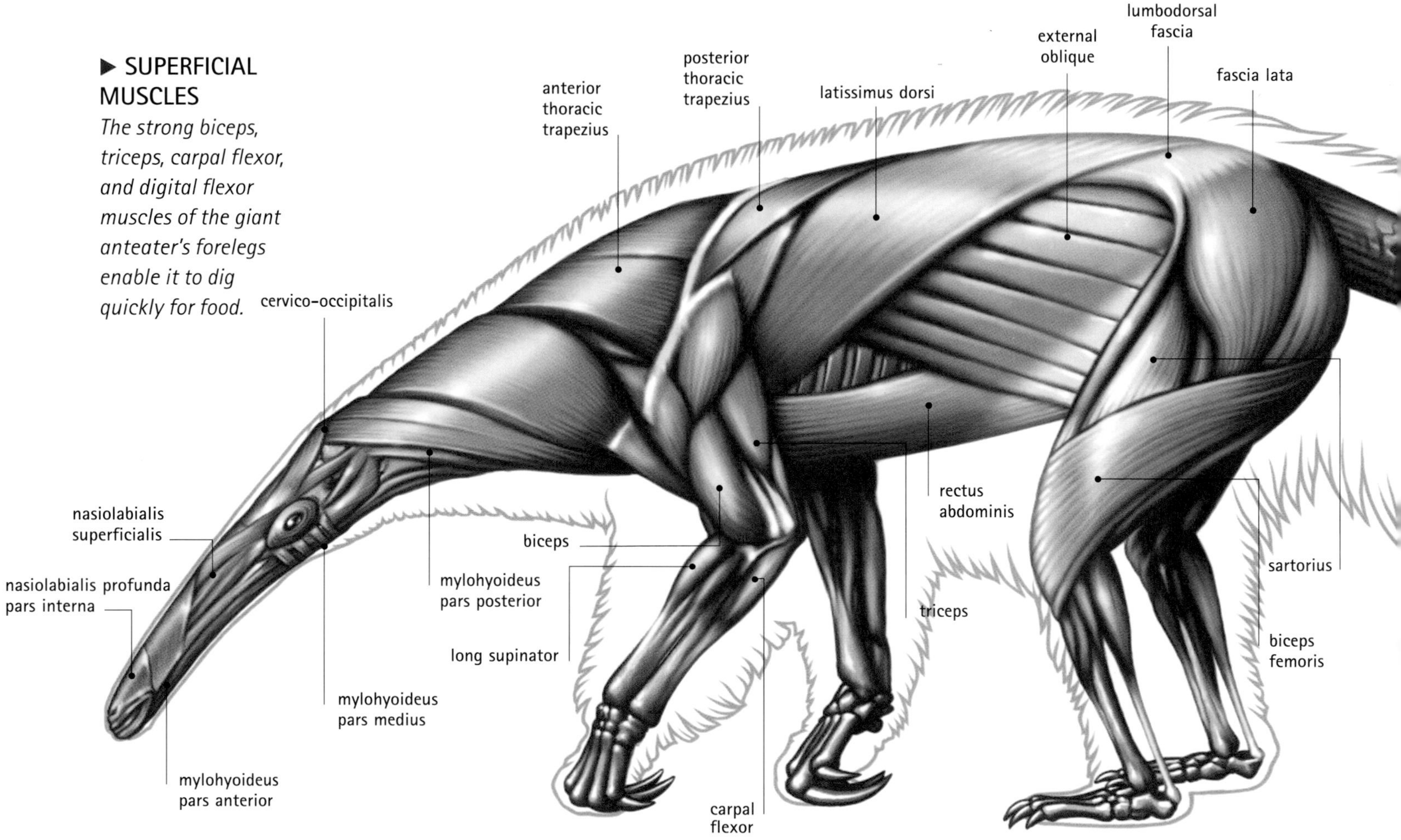

Most mammals, including humans, have a tongue that attaches to the throat. Rising up the throat from the sternum, an anteater's tongue comprises two parts, which merge below the base of the skull under the first neck vertebra. The combined tongue passes through the tubelike mouth and can extend up to 2 feet (61 cm) from the tip of the snout.

Getting food

The two sides of the lower jaw do not meet at the chin. Even though an anteater does not chew, it has masseter muscles attached to the bones of the lower jaw. In most mammals, the masseter muscles power chewing. In the giant anteater, masseter muscles have a different

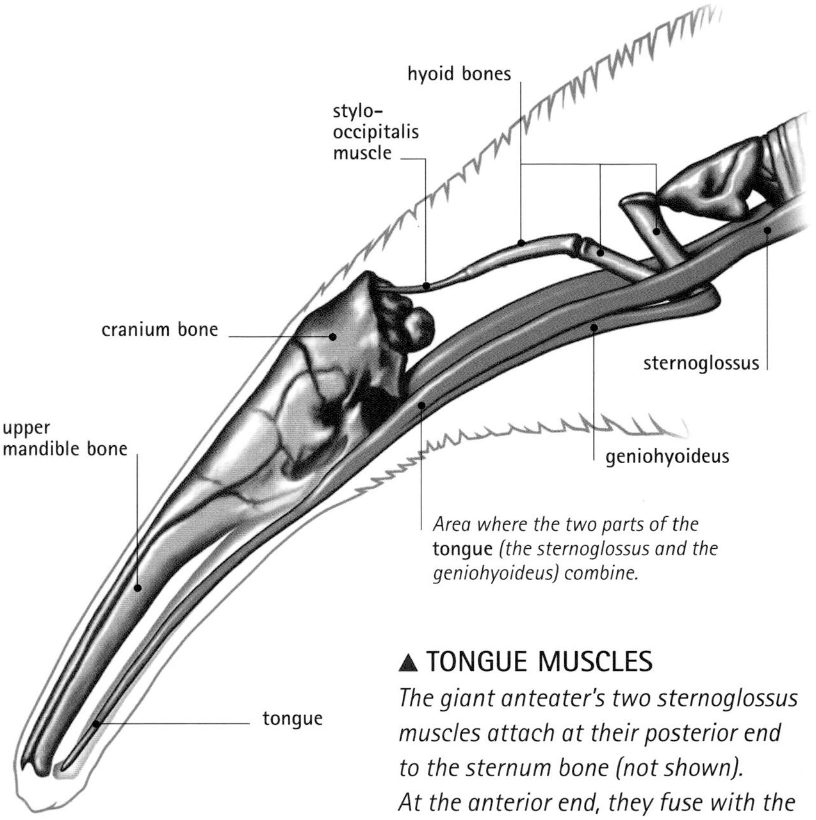

▲ TONGUE MUSCLES
The giant anteater's two sternoglossus muscles attach at their posterior end to the sternum bone (not shown). At the anterior end, they fuse with the geniohyoideus to form the single tongue.

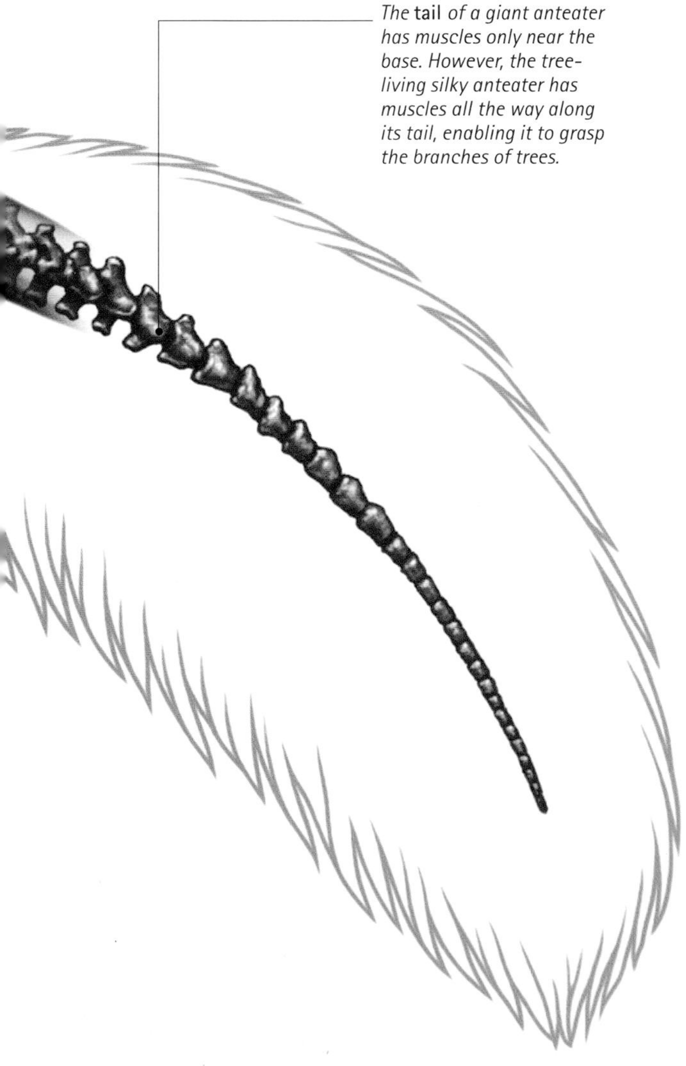

COMPARATIVE ANATOMY

A unique support

Many biological cylinders, such as the bodies of worms or sea cucumbers, or the tentacles of an octopus, do not rely on bones for support. Instead, they are supported by the pressure of fluids pushing outward from an internal cavity. This is called a hydraulic skeleton. For such a system to function, the cylinder needs a tough outer layer to resist the pressure, and the ability to bend to allow movement without crimping. To meet both these requirements, fibers just beneath the skin of these animals weave around the inside of the cylinder in a crossed helix shape, like a pair of spiral staircases. Biologists assumed that a very similar setup existed in penises, which, when inflated with fluid prior to copulation, form hydrostatic skeletons. However, recent research on the penis of armadillos, relatives of anteaters, has revealed an unexpected pattern of fibers. As in a worm, there is a set of fibers around the circumference of the penis and another set along its length, but there are no helical fibers. Unlike a flexible worm, a penis needs to be rigid when inflated and must avoid bending. The fiber structure discovered in the armadillo provides both rigidity and structural support. Similar arrangements probably occur in the penises of most other animals.

CLOSE-UP

Muscles of the forelimbs

A giant anteater's lower forelimbs are so specialized for breaking into termite hills and ants' nests that they make for difficult walking. The forelimbs are strong and support the weight of the animal. However, the picklike claws are so long that the animal cannot place its feet flat on the ground. A walking giant anteater stands on the outside of its wrists with its fore claws raised off the ground, tucked up behind the foot.

The forelimbs are heavily muscled and are attached to large shoulder, chest, and back muscles. These and the strong deltoids, biceps, and triceps of the upper forearm provide the muscle power of the arm. Strong muscles pass the length of the forearm to the wrist. A further long muscle, called the supinator, connects to a projection on the upper forelimb bone, the humerus, far above the elbow. The other end of the supinator forks into two, with one side attaching to each side of the wrist. The double attachment of the supinator enables the anteater to rotate its wrist.

Compared with the stout forearms, the anteater's back legs are slim and lightly muscled. The hind limbs are used not for digging, but only for walking and standing. The lighter muscles used in locomotion attach to lighter leg bones.

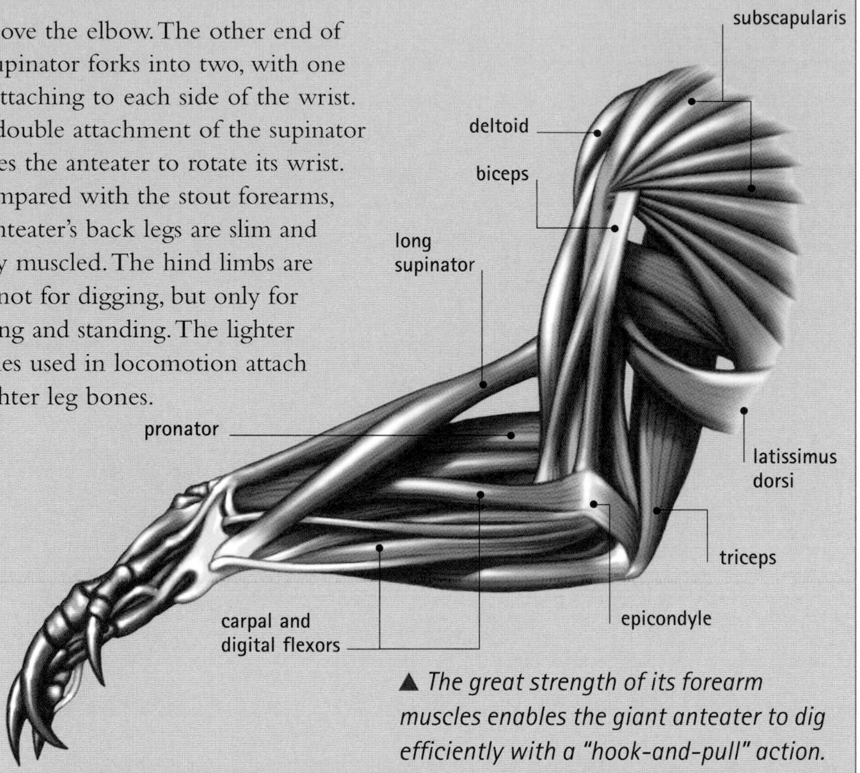

▲ *The great strength of its forearm muscles enables the giant anteater to dig efficiently with a "hook-and-pull" action.*

function. They rotate the two halves of the lower jaw in a movement that widens, then closes, the space for the long tongue to flick into and out of the mouth.

A sticky tongue

The tongue itself is cylindrical and sticky with saliva; its cylindrical shape makes it easier to take up ants, while the stickiness of the saliva helps keep the ants on the tongue as it is retracted into the mouth. The tongue is also armed with tiny backward-pointing barbs. The tongue barbs form from small raised areas of the tongue called papillae. The papillae stiffen when in use and further assist the anteater in drawing its food into the tiny mouth opening.

▶ Silky anteater

The muscles of a silky anteater's tail allow it to grasp the branch of a tree. This species also uses its hind feet and muscular, prehensile tail to clutch a branch when it adopts a defensive posture.

Nervous system

CONNECTIONS

COMPARE the very large olfactory organs of an anteater with the small olfactory organs of a primate such as a ***HUMAN***, for whom vision is far more important than smell.

COMPARE the poor eyesight of an anteater with the excellent vision of a predator such as an ***EAGLE***.

Vision and hearing are not a giant anteater's most important senses. Small eyes and poor vision do not impede the search for food, since ant and termite mounds are conspicuous and easy to find. Once the anteater has located a nest or mound, the ant or termite prey is close at hand, and smell becomes the most useful sense. A feeding anteater cannot see what is happening at the end of its snout well enough to select individual ants by sight.

The giant anteater's hearing is also rather limited. Giant anteaters do not need sensitive hearing because they locate their prey by other means and have little to fear from most predators. Unlike smaller mammals, large, well-defended mammals such as the giant anteater do not need sharp vision or large ears that can move in all directions. Smaller mammals need their ears and eyes to serve as early warning systems; giant anteaters do not have such a critical need to be on guard. The most frequent attacks on an anteater come from ants and termites. The small size of a giant anteater's external ears minimizes the surface area of skin exposed to ant stings.

IN FOCUS

Stinky messages

Anteaters rarely encounter others of their kind face to face. They range over very large areas and spend a lot of time resting or moving around only slowly. The chances of meeting a rival or mate are slim. As with many other mammals that range over wide areas, anteaters leave chemical messages around their territory. Anteaters deposit secretions from their anal glands as scent marks. These secretions can be formidably smelly. People have nicknamed tamanduas "stinkers of the forest" after their territorial anal secretions. Animals with an acute sense of smell such as anteaters can likely detect many things about the individual that left the scent. Information coded in chemical messages typically includes the sex of the animal and its readiness to mate.

▼ *The giant anteater's nervous system is similar to that of other mammals. The most noticeable difference, as shown in this schematic diagram, is the much larger olfactory lobes at the front of the brain.*

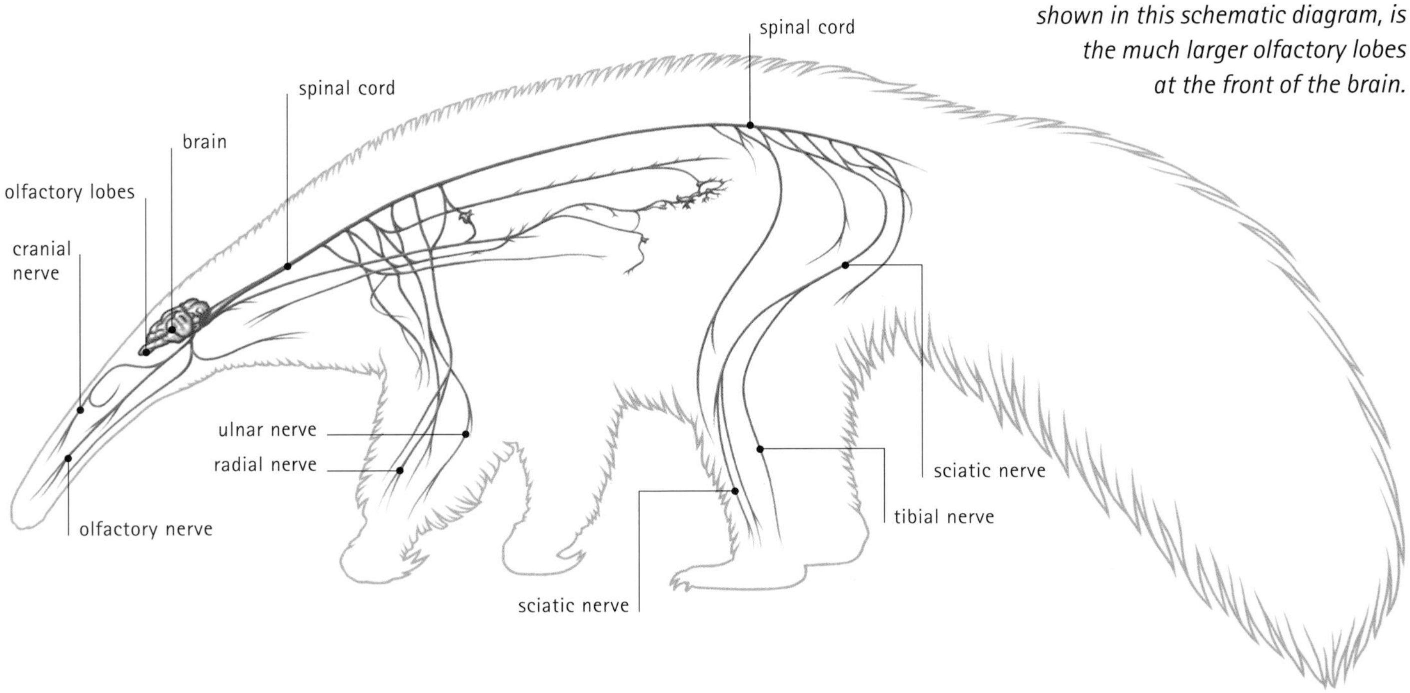

▶ *A hungry southern tamandua searches for tree-dwelling ants, termites, and bees. Like its relative, the giant anteater, the southern tamandua has large olfactory organs connected through nostrils at the tip of an elongated snout that enables it to detect the scent of the insects forming its diet.*

CLOSE-UP

The anteater brain

Mammals typically have a well-developed brain larger than that of animals in other vertebrate classes. However, in comparison with the brains of other mammals, an anteater's brain is very small for the animal's size. The giant anteater's acute sense of smell is reflected in the anatomy of its brain. The olfactory bulb and tracts are concerned directly with an animal's sense of smell. In animals that use vision more than smell, such as primates, the olfactory areas are tiny. Anteaters, armadillos, and aardvarks all have huge olfactory bulbs and tracts, and their sense of smell is among the most sensitive of all mammals.

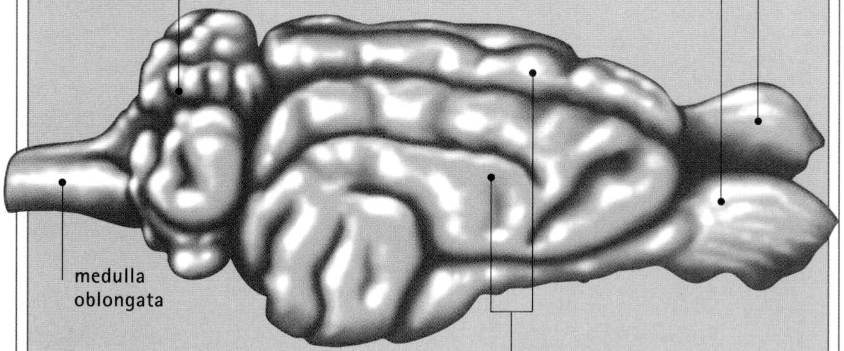
olfactory bulbs
cerebellum
medulla oblongata
The cortex is divided into the left and right cerebral hemispheres.

▲ BRAIN
Giant anteater

Sensitive sniffers

Different animals have differing needs for the sense of smell. Primates rely on their visual sense far more than their olfactory (smelling) sense, and they have very small olfactory organs. Aardvarks, armadillos, and anteaters all have very large olfactory organs relative to their size. This reflects the importance of smell to all these animals.

An anteater's nostrils are close to the end of its elongated snout. This is a typical position for the nostrils of animals that root around, sniffing out food. Other ant- or insect-eating mammals such as pangolins, aardvarks, hedgehogs, and echidnas also have nostrils at, on, or close to the end of an elongated snout. Pigs and elephants both have nostrils on the tip of their nose and use smell a great deal in searching for and gathering food.

Discriminating smells

The giant anteater can distinguish between different species of ants and termites just by their scent. That is an important skill for an anteater. Different species of ants have different defenses against predation. Using its extraordinary olfactory senses, the giant anteater can avoid those types of ants and termites with the most severe bites.

Circulatory and respiratory systems

Compared with most mammals, the xenarthrans have a low metabolic rate, meaning the body burns relatively little energy to fuel vital processes such as respiration. The heart pumps blood through arteries to keep tissues supplied with oxygen and nutrients. The blood returns through a system of veins. The giant anteater's regular heart rate is between 110 and 160 beats per minute. A mammal's heartbeat is an involuntary process. It is usually impossible for a mammal consciously to slow or quicken its heartbeat. The brain controls a mammal's heartbeat in response to the animal's level of activity, or likely activity if it is scared and preparing to flee.

COMPARATIVE ANATOMY

Chest measurements

The giant anteater has a small rib cage compared with other mammals of its size. Rib cage size gives clues to an animal's lifestyle. A large rib cage indicates that an animal needs large intakes of air. Whales need to take a huge breath before they dive because it may be a long time before they next have a chance to breathe. Cheetahs have a large chest to provide oxygen during their energy-sapping sprints after prey. Gazelles need large lungs to escape from such predators. Giant anteaters do not need to hold their breath or chase prey. Neither do they use speed or stamina to escape their enemies. Anteaters rarely run far or fast. Their most strenuous activity, on most days, is pulling down anthills. Small lungs and a small rib cage are perfectly adequate.

Breathing rates

A giant anteater breathes in and out between 10 and 30 times a minute. Whereas the mammalian heartbeat is involuntary, a mammal can raise or lower its rate of breathing consciously. Two actions combine to refresh the air in a mammal's lungs: the expansion or contraction of the rib cage and movement of the diaphragm. The ribs form a cage in the upper body that protects the lungs and heart from physical harm. The intercostal muscles crisscross the ribs. Together, they form the rib cage. The intercostals move the rib cage in and out as a mammal breathes. When the rib cage expands, the lungs expand with it. The expanding lungs suck in air. The diaphragm is a sheet of muscle below the lungs. The diaphragm pulls down when a mammal breathes in, further increasing the volume of the chest and lungs.

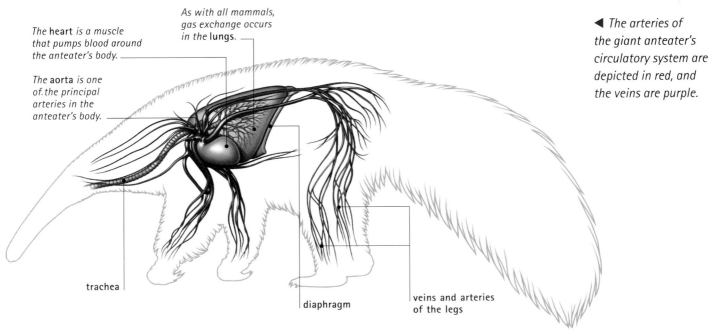

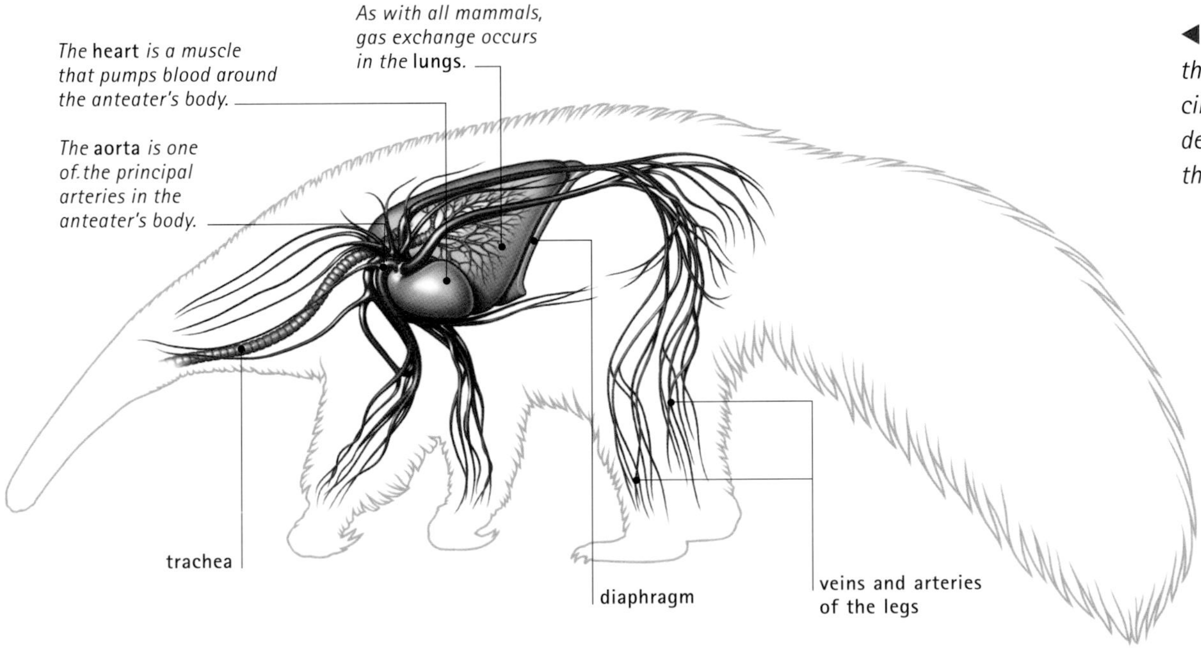

The heart is a muscle that pumps blood around the anteater's body.
As with all mammals, gas exchange occurs in the lungs.
The aorta is one of the principal arteries in the anteater's body.
trachea
diaphragm
veins and arteries of the legs

◀ *The arteries of the giant anteater's circulatory system are depicted in red, and the veins are purple.*

Digestive and excretory systems

Giant anteaters do not have teeth, and their jaws do not open wide like those of most other mammals. The giant anteater cannot chew. It sucks up and swallows as many ants as it can in the shortest possible time. The longer the anteater takes to eat its meal, the more time the defending ants or termites have to attack the anteater. By flicking its sticky saliva-covered and barbed tongue up to 2 feet (61 cm) from its mouth, the anteater can pick up ants with each flick. A giant anteater may extend its tongue as often as 150 times a minute. The speed of the tongue is vital for the anteater to gather the 30,000 or so insects it needs to eat each day. However, scientists have discovered that anteaters do not consume all the ants in the nests they predate. This behavior allows the ants to regenerate, and the anteaters' food base is conserved.

Constant swallowers

An anteater swallows almost continuously as it eats. It has very large salivary glands; their

COMPARATIVE ANATOMY

Different guts for different menus

In most mammals the structure of the intestines reflects diet and lifestyle. Meat eaters generally have shorter intestines than plant eaters. Omnivores such as humans eat both plant and animal food; they have mid-length intestines. Many herbivores have very long intestines or extra stomachs that enable fermentation of their food.

Anteaters eat ants and little else. They are most similar to meat eaters in intestinal structure. However, like other ant-eating mammals, giant anteaters have slightly longer intestines than most carnivores, and they also have a larger stomach. Many carnivores have a stretchy stomach. They are binge eaters, with occasional but large meals that need to be accommodated. A typical carnivore's stomach swells to accommodate a meal and shrinks after digestion, leaving the animal ready to hunt. Anteaters need plenty of room in their stomach to fit in meals of thousands of ants on a regular basis. Since they do not use speed or agility to catch their food, a large stomach does not affect their movement. Having a large stomach (relative to body size) would be a disadvantage for an animal relying on speed to hunt.

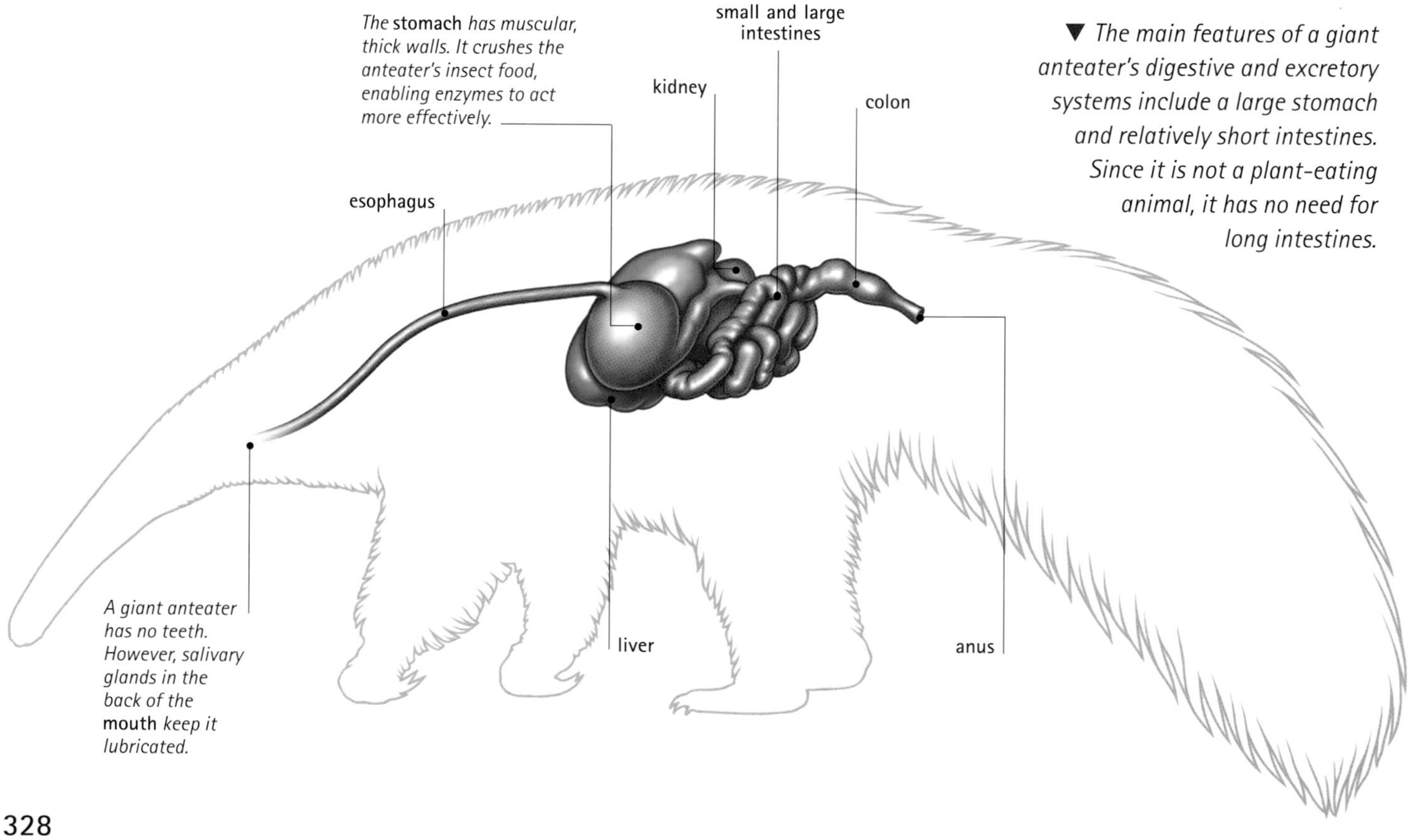

The stomach has muscular, thick walls. It crushes the anteater's insect food, enabling enzymes to act more effectively.
small and large intestines
kidney
colon
esophagus
A giant anteater has no teeth. However, salivary glands in the back of the mouth keep it lubricated.
liver
anus

▼ *The main features of a giant anteater's digestive and excretory systems include a large stomach and relatively short intestines. Since it is not a plant-eating animal, it has no need for long intestines.*

▶ *A giant anteater preys on ants in South America. Anteaters use their very long tongue to draw prey into their mouth. After passing along the esophagus, the ants are ground up in the anteater's muscular stomach.*

secretions lubricate the tongue as it passes backward and forward through the mouth. The saliva gives the tongue a sticky wet coating that picks up ants. The salivary glands are located behind the brain and open into the back of the mouth. Ducts pass down each side of the neck, ending at the sternum at the top of the chest. Astonishingly, a giant anteater's salivary glands are larger than its brain.

The giant anteater swallows its food without chewing. Instead, it has a powerful grinding stomach that breaks up its food. One species of pangolin has hard toothlike structures in its stomach that grind its food. A giant anteater's thick-walled grinding stomach can deal with similar food without such an armory of false teeth. The muscular stomach mashes up the insects, breaking open the hard exoskeletons and exposing the nutritious inner parts to the action of digestive enzymes.

Digesting food

Digestion is directed toward changing food into an absorbable form so it can enter the bloodstream and be taken up by the body's cells. Most digestion by enzymes takes place in the small intestine, and the products of digestion are absorbed across the walls of the small intestine. The digestive process is more efficient if the surface area through which this absorption takes place is large. Like those of other vertebrates, anteaters' intestines are coiled. Many other animals have longer intestines, but those of giant anteaters are nevertheless long in relation to body length. A covering of tiny, fingerlike villi and microvilli lines the inside of the digestive tract, increasing its surface area many times.

As with other mammals, solid waste products are passed from the intestines into the colon and rectum and excreted from the anus. Liquid waste, or urine, passes from the kidneys to the bladder for temporary storage. The female giant anteater differs from most other placental mammals in that she has a single chamber for her urinary and reproductive openings. This arrangement is more typical of marsupial mammals, birds, and reptiles than it is of placental mammals. A combined reproductive and urinary opening is called a cloaca. Beavers are another placental mammal group in which females have a cloaca.

Reproductive system

CONNECTIONS

COMPARE the precocious newly born young of a giant anteater with the young of a marsupial such as a ***KANGAROO***. Kangaroo young are born in an embryonic state.

COMPARE the single opening for the urinary and the reproductive tract of the female giant anteater with the single opening, or cloaca, found in a bird such as an ***ALBATROSS***.

Giant anteaters are placental mammals. Placental mammals nourish their unborn young in the uterus. If, after fertilization, the embryo implants into the wall of the uterus it may grow into a fetus. A temporary organ called the placenta then grows out between the uterus wall and the embryo itself. The placenta is packed with blood vessels that carry food and oxygen from the mother directly to the growing fetus through the umbilical cord.

Unusual reproductive organs

Xenarthrans, including the giant anteater, have reproductive organs that differ greatly from those of other mammals. The female has a single opening for her reproductive and urinary tract, an arrangement that is common in birds and reptiles but unusual in placental mammals. The beaver is one of the few other mammals with a combined urinary and reproductive tract in the female. The anus is separate from the reproductive and urinary tract.

▼ *When giving birth, a female anteater stands upright on her hind legs and uses the strong bones and muscles in her tail to provide a third support.*

When mating occurs

Giant anteaters may breed at any time of year in captivity. Scientists believe that in the wild adult male and female giant anteaters come together only to mate. They most often mate in fall and give birth, usually to a single offspring, in spring. The duration of pregnancy in animals is called the gestation period. An anteater fetus grows to a birth weight of around 3 pounds (1.3 kg) in a gestation time of around 190 days. However, this time is variable. The shortest recorded gestation of a giant anteater was 142 days. No one knows why the gestation period varies so much. Perhaps the timing depends on the mother's food intake or maybe longer gestation periods indicate that females can delay development of the embryo, as badgers can.

COMPARATIVE ANATOMY

Where are the testes?

Male mammals produce sperm in their testes. In many mammals, the testes hang outside the body. External testes may enable an animal to regulate their temperature and so ensure that the optimal temperature for sperm production is maintained. Some mammals, particularly primates, use their testes as a badge. Some monkeys have brightly colored testes that they display during social interactions. Such performances are impossible for a male giant anteater. That is because his testes are hidden inside his body. The testes of all xenarthrans lie in the abdomen, between the bladder and the anus.

▲ *Soon after birth the baby anteater clings to its mother's back, holding on to the coarse hair with its long claws.*

A standing start

Giant anteaters give birth standing up, using the tail as a prop. This makes birth easier, since gravity lends a hand. Baby anteaters are termed precocious because they are born in a relatively advanced state. A newborn giant anteater has fur, and even shows the characteristic dark flank stripe. Almost right away it has the strength in its forearms to hang from the mother's hair or ride on her back. When it clambers aboard, the female licks the baby clean. A young anteater is born sightless and opens its eyes after six days. For the first two to six months of its life the young anteater suckles milk from its mother. Female xenarthrans have nipples near the armpits, or on the chest or abdomen. A female anteater may be ready to mate again just six or seven months after giving birth.

A young anteater rides on its mother's back for six to nine months, although it is able to run after a month. The powerful forearms take a long time to grow fully and enable the anteater to break into large anthills. The youngster, therefore, feeds alongside the mother for two years before becoming independent. A young anteater is not sexually mature until it is three or four years old.

JOHN JACKSON

IN FOCUS

Growth in young anteaters

The snout of a young giant anteater grows faster than the rest of the animal. A newborn anteater has a relatively short snout, but as the animal grows, its nose becomes ever longer in proportion to its body: up to 18 inches (46 cm) in an adult. Such a pattern of uneven growth is called allometric growth.

FURTHER READING AND RESEARCH

Macdonald, David. 2006. *The Encyclopedia of Mammals.* Facts On File: New York.

Animal Diversity Web: animaldiversity.ummz.umich.edu/site/accounts/information/Xenarthra.html

Giant centipede

ORDER: Scolopendromorpha FAMILY: Scolopendridae
GENUS: *Scolopendra* and 7 others

Centipedes are fierce predators that live in many terrestrial habitats around the world. Most occur in moist tropical areas, but some live in arid semideserts, and others prosper in cold subpolar regions. One species lives on the seashore and can survive for long periods underwater.

Anatomy and taxonomy

Scientists group all organisms into taxonomic groups based largely on anatomical features. Giant centipedes belong to the order Scolopendromorpha, one of three orders of centipedes. Each order comprises one or more families, and each family is made up of several genera and species.

• **Animals** Animals are multicellular organisms with cells that are generally organized into tissues and organs. Both animal and plant cells contain membrane-bound structures called organelles (mini-organs). Unlike plants, however, which make their own food, animals must eat to get energy. Almost all animals have guts that enable digestion, and most animals have a nervous system that controls how they interact with their surroundings.

• **Arthropods** Arthropoda is the largest and most successful phylum of organisms on Earth, yet despite its vast diversity the general body plan of arthropods is relatively constant. They have segmented bodies, although segments are often fused to form units such as the head and abdomen. Arthropods have a tough outer "skin" called an exoskeleton. It protects the internal organs and serves as an attachment point for muscles. To grow, an arthropod must shed, or molt, its exoskeleton.

Arthropods have pairs of jointed appendages, such as legs, mouthparts, and antennae. Internally, all arthropods have a ventral (running along the underside) nerve cord and a dorsal (running near the top surface) blood vessel. The rear part of the dorsal vessel pumps liquid called hemolymph around the body cavity (or hemocoel).

• **Uniramians** Legs and other appendages of uniramians form a single, unbranched structure. Other arthropods, such as crustaceans, have biramous appendages—their legs, for example, have an outer branch, which often forms a gill, and an inner branch that is used for walking.

• **Myriapods** Among the most ancient of all terrestrial animal groups, myriapods have a body that is divided into two main sections: the head and the trunk. The trunk is

▶ *This tree shows the major groups to which centipedes belong, based on recent structural and genetic research; one small order, the Craterostigmomorpha, has been omitted for clarity. Higher-level relationships are controversial. Biologists are unsure whether myriapods are more closely related to hexapods (as here) or chelicerates, the subphylum that contains spiders.*

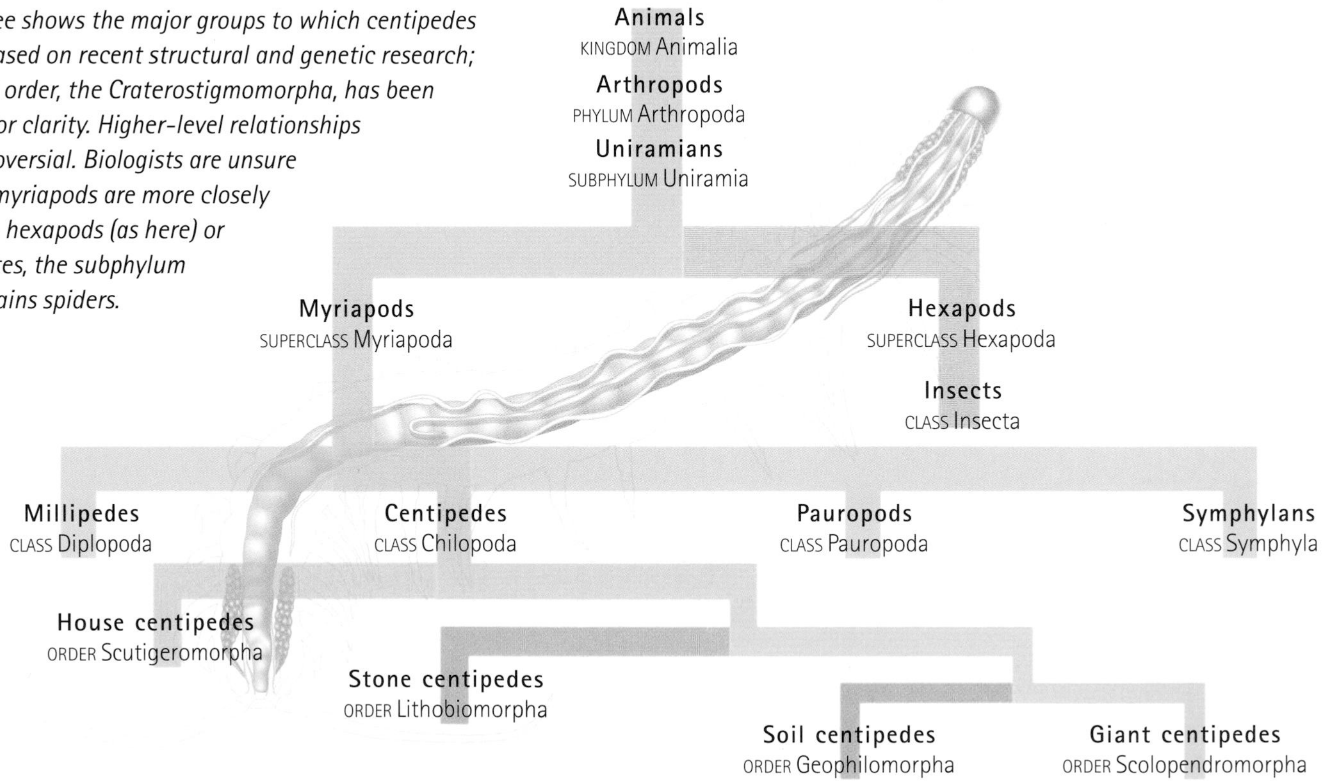

segmented, with most segments bearing one (or, in millipedes and pauropods, two) pairs of legs. Most myriapods have glands on each segment that release defensive chemicals, and a specialized sensory structure called the Tömösvary organ on the head. Internally, myriapods are metameric (structures are repeated along the length of the body).

Myriapods must molt to grow. Most increase their length by one segment at each molt, though some centipedes hatch from their eggs with the same number of segments as the adults. Like insects, myriapods breathe through a tracheal system; increasingly fine tubes carry oxygen to the tissues where it is needed. However, myriapods cannot close their spiracles—holes in the exoskeleton through which gases pass—so they are susceptible to water loss.

• Pauropods and symphylans These classes (major groups) of myriapods contain tiny, soil-dwelling animals. Symphylans resemble small centipedes, with one pair of legs on most of the trunk segments. However, symphylans eat plant material and have an appendage at the rear tip of the body called a spinneret with which they can spin silken cocoons. Pauropods feast on fungal threads and decaying material in the soil, which they detect using their unusual, forked antennae. Pauropods have pairs of fused segments, each bearing two pairs of legs.

• Millipedes With around 10,000 species, the millipedes are the largest class of myriapods. The first segment behind a millipede's head (the collum) is legless, as are the last

FEATURED SYSTEMS

EXTERNAL ANATOMY Centipedes are long, segmented animals with relatively long legs adapted for fast running. *See pages 334–335.*

INTERNAL ANATOMY Centipedes have a typical arthropod internal anatomy with many metameric (repeated) structures. *See page 336.*

NERVOUS SYSTEM The brain connects to sense organs such as the sensitive antennae, the ocelli, and the Tömösvary organ. *See page 337.*

CIRCULATORY AND RESPIRATORY SYSTEMS Hemolymph is pumped around the body by the heart; this substance is especially important in house centipedes, which use a respiratory pigment to transport oxygen around the body. *See page 338.*

DIGESTIVE AND EXCRETORY SYSTEMS Prey is immobilized with venom from the maxillipeds, fanglike structures unique to centipedes. *See pages 339–340.*

REPRODUCTIVE SYSTEM Sperm is transferred from the male to the female in packets following a complex courtship. *See page 341.*

▲ *A giant desert centipede is able to walk quickly along the branch of a tree in search of prey or to escape a predator.*

few segments near the tip of the trunk. The three segments behind the collum bear one pair of legs each, but all the other segments have two pairs. Millipedes generally have more legs than other myriapods—up to 800. Millipedes have mouthparts called mandibles with which they chew decaying plant matter. All millipedes have segmented antennae tipped with sensitive cells, with the horseshoe-shaped Tömösvary organ at the base. Many millipedes have clusters of light-sensitive ocelli on their head.

• Centipedes Unlike millipedes, all centipedes are fast-moving hunters. Their first six segments are fused and form the head. The legs of these segments form mouthparts, including a lower lip, two pairs of long maxillae used to manipulate food, and a pair of clawlike maxillipeds, or fangs. The tips of these sharp fangs are connected via a duct to venom glands inside. The maxillipeds pierce the body of prey and inject venom.

The rest of the body, the trunk, consists of unfused segments. In centipedes, unlike millipedes, each segment bears just one claw-tipped pair of legs—between 15 and 177 pairs in total. Most walk on the tip of the claw. The tarsi of house centipedes form a longer "foot." Centipedes do not have compound eyes like insects. Many are blind, and the remainder have only simple ocelli. The antennae, which attach to the head, are much more important for detecting prey, mates, and enemies.

External anatomy

Like an insect's, a centipede's body is divided into distinct sections. Insects have three tagmata, or sections, but centipedes and other myriapods have only two: the head and trunk. Covering both sections is an external skeleton, or exoskeleton, which contains a tough protein called chitin that gives it strength. The exoskeleton protects the centipede's internal organs and serves as an attachment point for the muscles.

Water wasters

A centipede's exoskeleton is tough, but it differs from an insect's in one key way: it lacks a waxy outer cuticle that keeps water inside. This absence leaves centipedes vulnerable to water loss. Further water is lost through the spiracles (air holes) that, in most species of centipedes, cannot be closed. Centipedes are forced to live in humid environments, or to conserve water through behavioral means—by foraging in the cool of the night or by living under rotting logs or in soil, for example.

The exoskeleton is rigid and cannot expand. Thus, a centipede must molt its exoskeleton to grow. Molting in many species also leads to an increase (by one) of segments and (by two) of legs. When the centipede is ready to molt, a new exoskeleton grows beneath the old one. The centipede then stops foraging and retreats to a sheltered place. A thin layer of fluid forms between the old exoskeleton and the new. Using the pressure of hemolymph (circulatory fluid) to swell its body, the centipede splits its old exoskeleton between the first and second trunk segments. The animal then climbs free, swiftly expands, and waits for the new exoskeleton to harden. The first meal for a newly molted centipede is usually its old exoskeleton.

Centipedes bear a range of appendages. There are a pair of long antennae and several separate mouthparts on the head, including two pairs of maxillae. Just past the head are the maxillipeds, modified legs used as venom-injecting fangs. Each of the rest of the body segments (with the exception of the last two) bears a pair of legs. The last pair are much longer than the rest; those of giant centipedes bear sharp spines and are used for defense, in combination with chemical discharges.

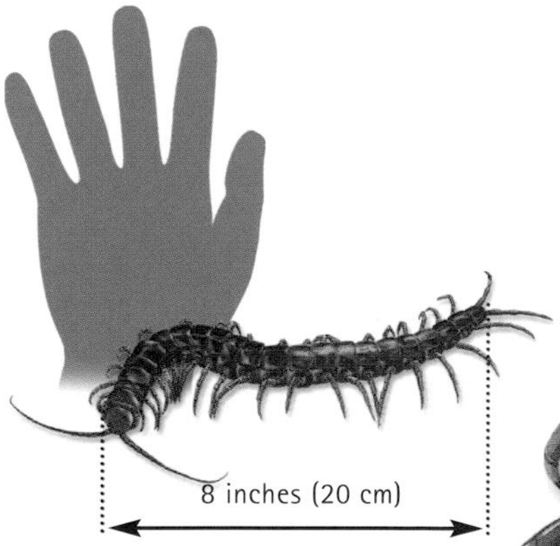

▼ Giant redheaded centipede
All centipedes have a segmented body and paired legs. The giant redheaded centipede has either 21 or 23 pairs of legs. The legs of most species increase slightly in length along the body. At high speeds, the longer legs step up and over the shorter legs in front of them to avoid tripping.

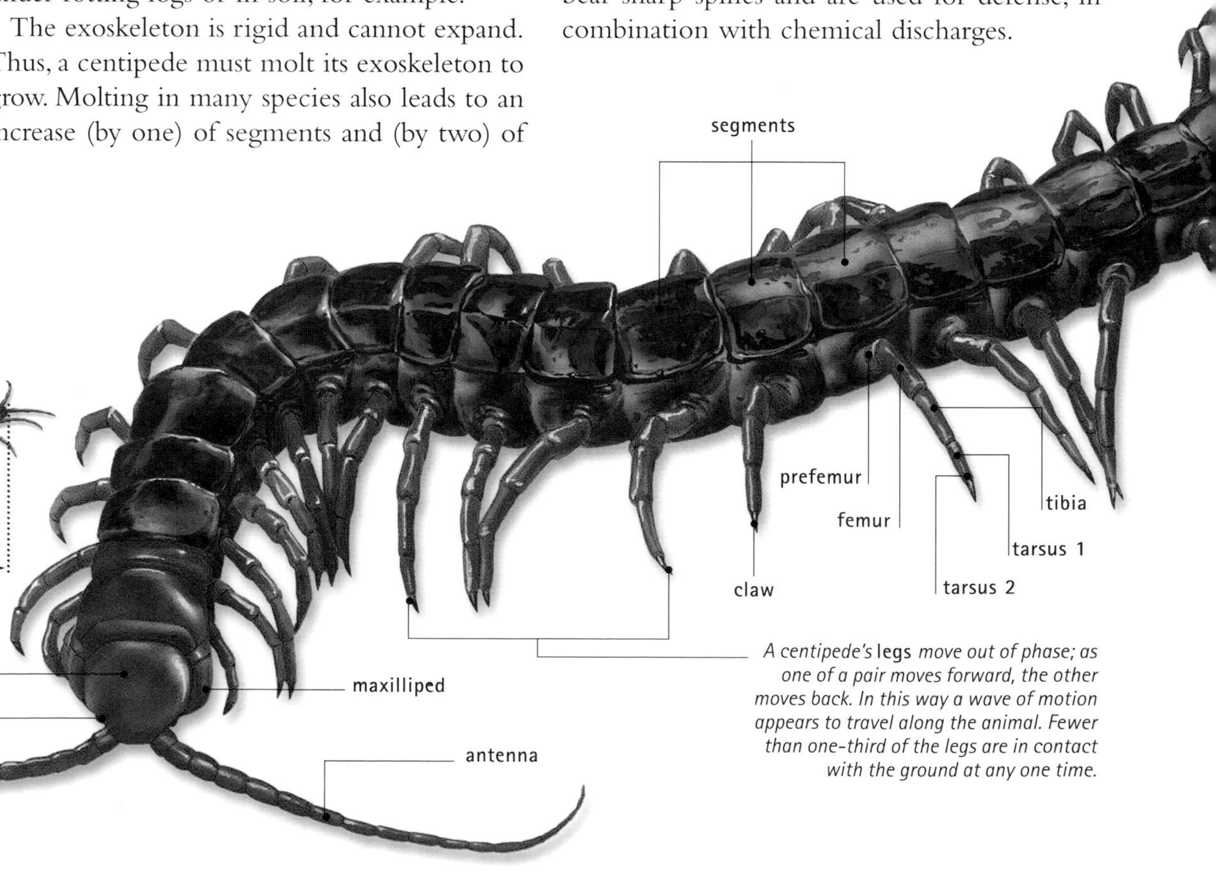

A centipede's legs *move out of phase; as one of a pair moves forward, the other moves back. In this way a wave of motion appears to travel along the animal. Fewer than one-third of the legs are in contact with the ground at any one time.*

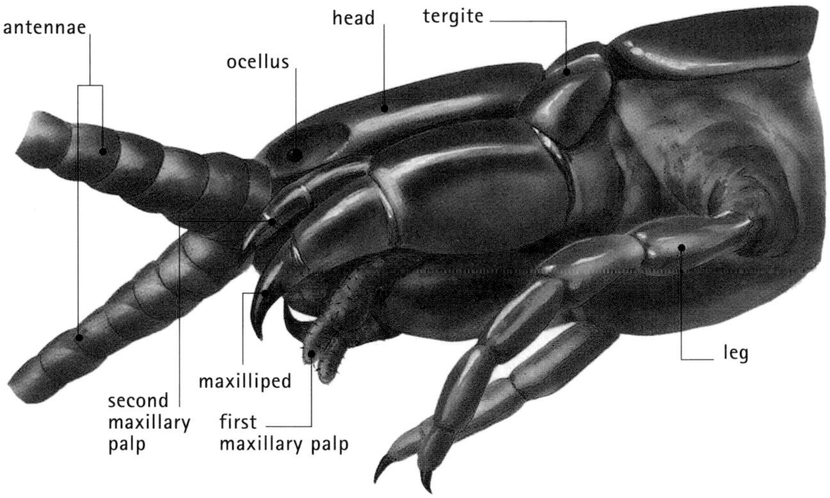

▲ **HEAD PROFILE**
Giant redheaded centipede
A centipede, like other arthropods, has jointed mouthparts, antennae, and legs.

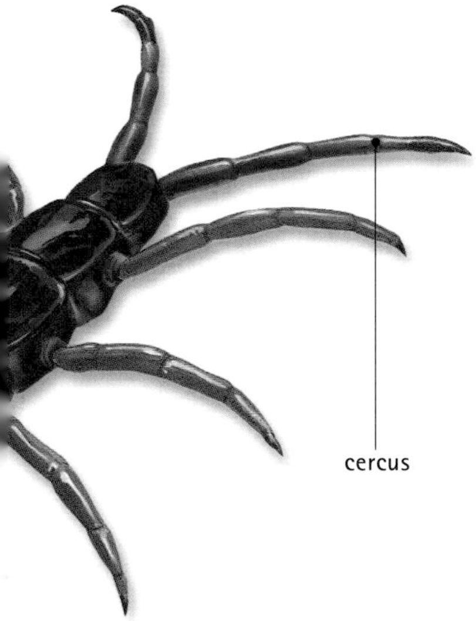

Leg anatomy

Centipedes' legs are divided into several segments and attach at the coxa with a flexible joint to the body. Most centipedes walk "on tiptoe," with the tips of the tarsi in contact with the ground. House centipedes' tarsi are multisegmented, so these centipedes can walk on their "feet." With far more of each leg in contact with the ground, house centipedes' legs generate more power, allowing the animals to run very quickly—they can motor along at 16.5 inches (42 cm) per second.

Fast-running centipedes tend to have long legs that can move through a large angle. By contrast, most millipedes are burrowers that bulldoze their way through soil. They have shorter legs and much slower movements. To reach high speeds, centipedes increase their stride length. They do this by sending waves of undulation along the trunk in time with the movements of the legs. Each leg is connected to neighboring legs by a band of elastic material that helps the animal coordinate movement. Running fast, however, can cause problems. The legs of most species increase slightly in length along the body. At high speeds, the longer legs step up and over the shorter legs in front of them to avoid tripping over each other.

EVOLUTION

Legs and climate

Leg number varies enormously among centipedes. Some have as few as 30, others more than 350. In 2000 biologists showed that the number of legs can vary within a single species, too. The researchers compared *Strigamia maritima* centipedes from southern England with specimens from the north of Scotland, some 450 miles (720 km) away. They found that southern centipedes have, on average, two segments (and four legs) more than their northern cousins. The difference is probably due to climate; in the colder north there is less food available, so the centipedes devote fewer resources to growth. Such differences within a species may, over long periods of time, lead to the separation of one species into two. Such separation is called speciation.

COMPARATIVE ANATOMY

One becomes two

Centipedes have just one pair of legs on each segment, but millipedes and their tiny cousins, pauropods, have two pairs on most of their segments. Why is that? Millipedes are divided into sections called diplosegments. Each represents the fusion of two separate segments into one. Hundreds of legs pushing together to drive a millipede forward may cause the trunk to buckle, but increased structural support from the diplosegments helps resist this. So millipedes and pauropods can burrow more easily than centipedes.

A pauropod has two pairs of legs on each segment.

Internal anatomy

CONNECTIONS

COMPARE the metamerism, or repeating internal structures, of a giant centipede with those of an ***EARTHWORM***.

COMPARE the double-stranded centipede ventral nerve cord with the fused double strand of an insect such as a ***DRAGONFLY***.

Internally, centipedes are similar to other arthropods, although they show a greater degree of metamerism (the repetition of structures) than most. Structures common to most arthropods include ostia (slits) in the heart, ganglia (nerve bundles), and spiracles (respiratory openings).

Centipedes have an internal body cavity, or hemocoel, through which hemolymph is pushed by the heart; the heart lies close to the dorsal (top) surface of the centipede. Smaller, noncontractile vessels run along the underside of the animal and in the head. Waste products in the hemolymph are filtered out by a long pair of structures called Malpighian tubules.

Guts and gonads

The gut runs through the center of the centipede from mouth to anus. Malpighian tubules pick up waste products from the hemocoel and empty them into the gut. Below the gut is the ventral nerve cord, which reaches from the brain to the tip of the trunk.

Centipedes are dioecious animals—they have two sexes and reproduce sexually through the fusion of sperm and egg. The sex organs (gonads) do not occur in pairs and lie near the end of the body. Sperm is passed from a male centipede to a female inside a packet called a spermatophore. The eggs are fertilized during oviposition (egg laying). A system of tubes called tracheae runs through the centipede's body. The tracheae connect to the outside through pores in the exoskeleton called spiracles. The tracheae take oxygen directly to the tissues where it is needed.

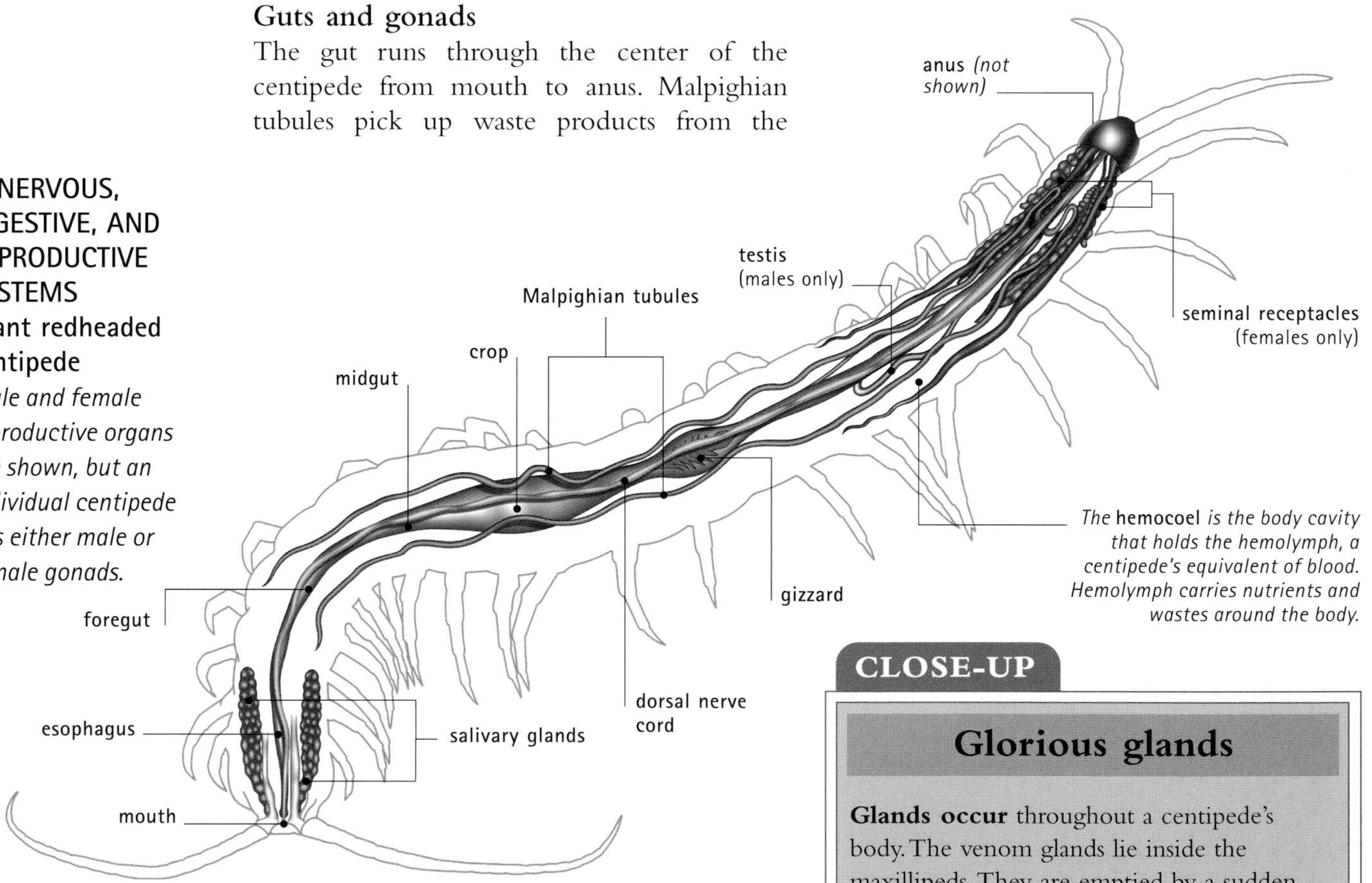

▶ **NERVOUS, DIGESTIVE, AND REPRODUCTIVE SYSTEMS**
Giant redheaded centipede
Male and female reproductive organs are shown, but an individual centipede has either male or female gonads.

CLOSE-UP

Glorious glands

Glands occur throughout a centipede's body. The venom glands lie inside the maxillipeds. They are emptied by a sudden muscular contraction. Other glands release their contents more slowly. The salivary glands, for example, dribble saliva out through the mouth and onto food.

Nervous system

Like those of other arthropods, centipedes' brains are composed of three distinct sections. The deutocerebrum is an elongated, branched structure linked to the two antennae. The deutocerebrum joins to the tritocerebrum, beneath which runs the protocerebrum, which is linked to the eyes. The brain lies above the esophagus, around which nerves skirt to connect the tritocerebrum to a knot of nerve fibers called the subesophageal ganglion. Two long strands of nervous tissue extend from this ganglion. The strands form the ventral nerve cord.

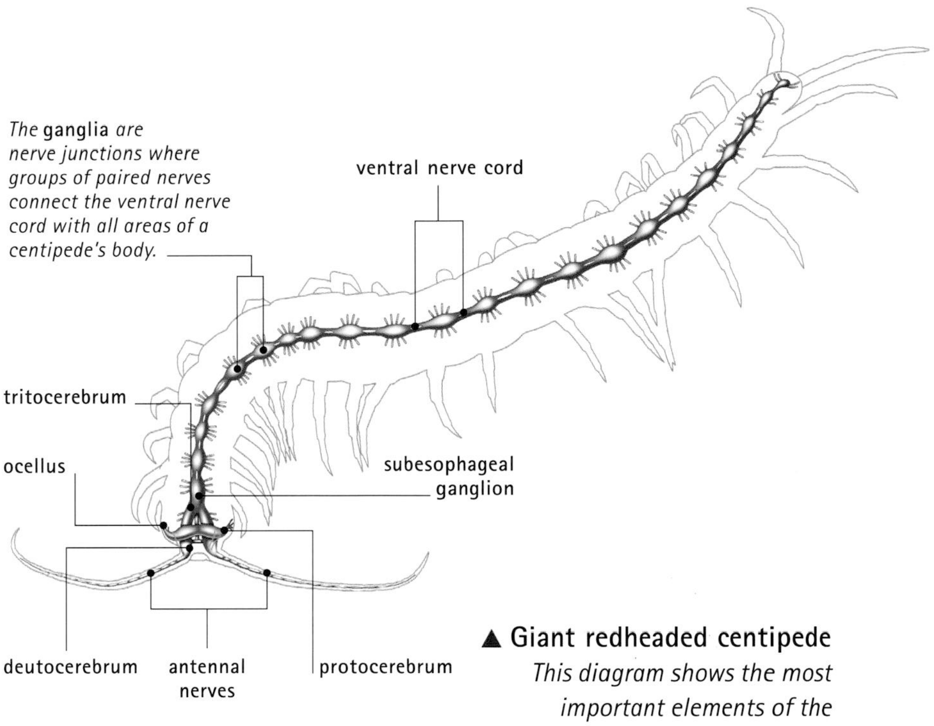

The ganglia are nerve junctions where groups of paired nerves connect the ventral nerve cord with all areas of a centipede's body.
ventral nerve cord
tritocerebrum
ocellus
subesophageal ganglion
deutocerebrum
antennal nerves
protocerebrum

▲ **Giant redheaded centipede**
This diagram shows the most important elements of the centipede's nervous system. Many small nerves branch from the ganglia and service various parts of the animal's body.

Nodules in the nerve cord

The nerve cord is a metameric structure—elements repeat in each segment along the body. The subesophageal ganglion is made of fused ganglia from several distinct segments. They joined to form the head in the ancient ancestors of centipedes. The next ganglion controls the maxillipeds. Each subsequent trunk segment has a ganglion with four or five major nerves leading from it. These nerves control the movement of the legs and many other processes.

Senses

Most centipedes have rather poor eyesight, but their long antennae are richly supplied with hairs that are extremely sensitive to touch and chemical stimuli. Centipedes have many touch receptors (movable bristles, or setae, on their legs, head, and body), which are connected to sensory neurons, or nerve cells. Information about the environment passes from the receptors to the centipedes' brain, and instructions are passed back from the brain. Most centipedes and other myriapods also bear a structure at the base of each antenna called the Tömösvary organ. Many sensory neurons converge there. The precise function of the Tömösvary organ is not known, but it probably allows the centipede to detect vibrations in the air or traveling through the ground. However, the Tömösvary organ is not present in giant centipedes.

COMPARATIVE ANATOMY

Comparing arthropod eyes

Most uniramian arthropods develop simple eyes, or ocelli, at some stage in their lives. Many insects develop sophisticated compound eyes, each containing thousands of individual units by the time they reach adulthood; others already have compound eyes in the nymphal stage. Centipedes and other myriapods have only ocelli; a centipede sees the world around it solely in terms of light and dark. That is not surprising, since most centipedes live in soil or are active at night and have little need for sharp vision. Many groups of myriapods have no eyes. However, house centipedes are fast-running predators that hunt above ground; they need vision more than most. House centipedes have clusters of around 200 ocelli on each side of the head. They form a structure similar to a true compound eye. Although less effective than an insect eye, house centipedes' ocelli represent an example of convergent evolution. This is the evolution of similar structures in unrelated groups of organisms in response to similar environmental challenges.

Circulatory and respiratory systems

Centipedes have an "open" circulation system with two main vessels, one dorsal (running close to the top surface of the animal) and one ventral (running near its underside). Hemolymph, the centipede's equivalent of blood, is pumped around the body by the long, tubular dorsal vessel, or heart. The heart extends from the last leg-bearing segment to the first trunk segment, where it is linked to the ventral vessel by a short branch called the maxilliped arch.

Both main vessels have extensions that reach into the head. Arteries leave these extensions to supply different head structures. The antennae and mandibles are supplied by arteries from the aorta, the front part of the dorsal vessel. The maxillae, however, are supplied by arteries extending from the ventral vessel.

Smaller, open-ended vessels branch from the main vessels. Hemolymph is pumped through these before passing into the hemocoel, or body cavity. Hemolymph moves back into the heart through channels called ostia, of which centipedes have a pair in each segment.

Transporting oxygen

Most centipedes get oxygen to their tissues by a system of tubes called tracheae. The tracheae open at the spiracles, circular pores in the exoskeleton, which are just above the legs on most segments. Several tracheae branch from each spiracle; dust-catching hairs surround the entrance to each. The tracheae themselves are strengthened by spirals of tough fibers and look like vacuum cleaner hoses. Most centipedes are unable to close their spiracles. Since centipedes are active mostly when it is cool and dark, water loss through the pores is not great. However, at least one species of giant centipede can close its spiracles.

Only house centipedes rely, as humans do, on blood to take oxygen around the body. They have just seven spiracles, each of which opens into a small sac close to the heart. This sac acts as a lung. Five tracheae inside the sac allow air to come into contact with the hemolymph, which contains a pigment called hemocyanin. Oxygen binds to the hemocyanin and is carried to the cells that need it. House centipedes have a thick heart, which maintains a steady flow of blood, with additional pumping organs in the head.

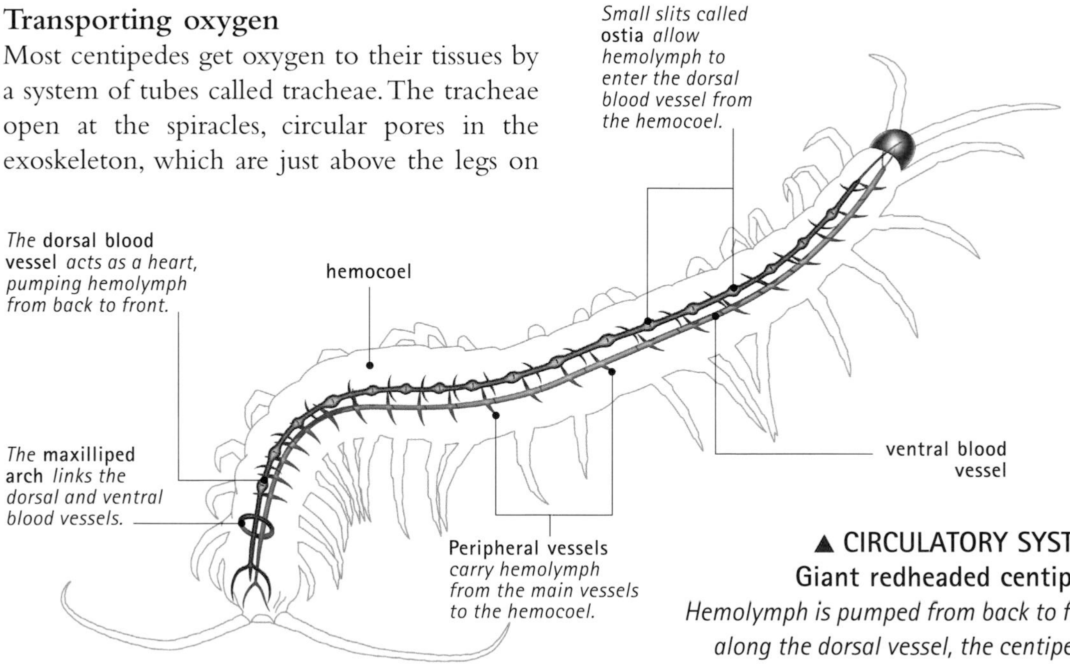

Small slits called ostia allow hemolymph to enter the dorsal blood vessel from the hemocoel.
The dorsal blood vessel acts as a heart, pumping hemolymph from back to front.
hemocoel
ventral blood vessel
The maxilliped arch links the dorsal and ventral blood vessels.
Peripheral vessels carry hemolymph from the main vessels to the hemocoel.

▲ **CIRCULATORY SYSTEM**
Giant redheaded centipede
Hemolymph is pumped from back to front along the dorsal vessel, the centipede's heart. Hemolymph passes into the hemocoel via peripheral vessels, then back into the heart via slits called ostia.

Digestive and excretory systems

CONNECTIONS

COMPARE the excretory system of a centipede with that of an insect such as a ***HOUSEFLY*** or ***WEEVIL***.

COMPARE the maxillipeds of a centipede with the fangs of a ***TARANTULA***.

Centipedes rely on powerful venom to immobilize their prey. The venom is administered through the maxillipeds, a pair of highly modified legs tipped with sharp, hollow fangs that extend from just behind the head. The venom is produced by glands inside the base of each maxilliped and is injected through ducts when the maxillipeds are forced into a victim's tissues.

Terrible toxins

Centipede venom contains a complex cocktail of chemicals that blocks the prey's nervous system. Giant centipedes' venom also contains cytolysins, chemicals that break down and destroy cells. This toxic arsenal quickly paralyzes the prey; the maxillipeds and second maxillae pin the prey down while the venom takes effect.

A set of glands called the salivary glands open near a centipede's mouth. Saliva is secreted onto food, helping to soften and partly digest it. The mandibles and first maxillae are then used to bite off small chunks that pass into the mouth, along the long esophagus, and into the crop.

The crop acts as a food storage area. Food passes from the crop to the gizzard, a muscular section of gut that crushes food. The gizzard contains forward-pointing spines that help sieve out large particles. A ring of muscle controls the movement of food into the midgut. There, food is coated by a membrane called the peritrophic envelope. Enzymes pass

COMPARATIVE ANATOMY

How millipedes feed

Almost all millipedes feed on plants or dead, decaying material. They bite off large pieces of material with their mandibles, mixing it with secretions from the salivary glands. The mandibles chew the food a little before it passes into the mouth. A few species feed on plant sap. Their mouthparts are modified to form a sharp beak that pierces the sap-carrying veins of plants.

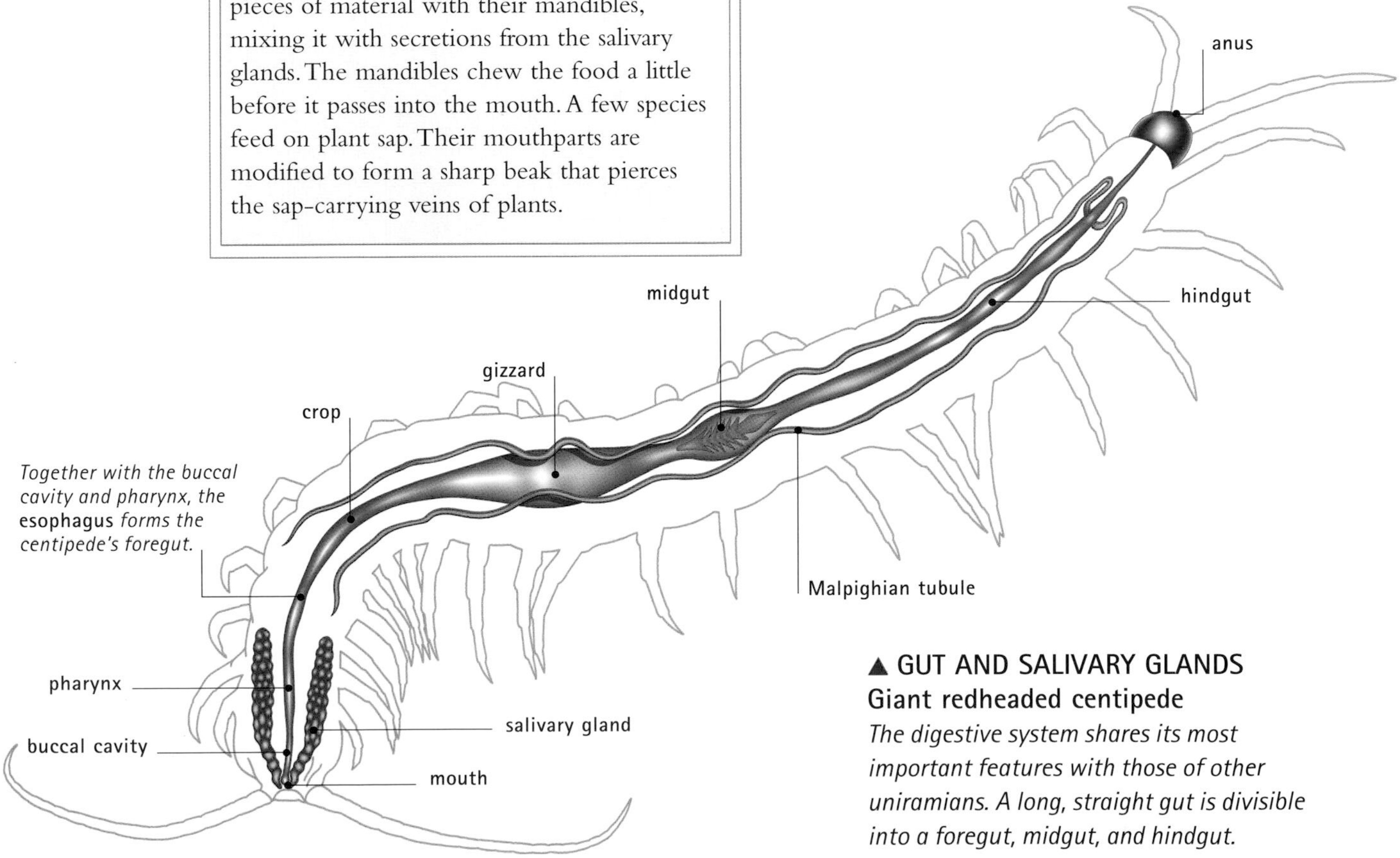

▲ GUT AND SALIVARY GLANDS
Giant redheaded centipede
The digestive system shares its most important features with those of other uniramians. A long, straight gut is divisible into a foregut, midgut, and hindgut.

through pores in the envelope, and digestion takes place. Food is absorbed through the walls of the midgut, from where it passes into the hemolymph. Undigested food passes through the hindgut and leaves the body at the anus, which opens at the last body segment.

Waste management

Like insects, centipedes get rid of waste from the body with ducts called Malpighian tubules. Centipedes have just one, very long pair of these structures. Waste moves into the tubules from the hemocoel, and passes into the midgut or hindgut. With their lack of a waterproof cuticle, and open spiracles, centipedes must avoid wasting water at all costs. Much of the animal's waste is converted to ammonia. This chemical can be released without the extra water loss required to flush away some other nitrogenous wastes, such as urea in humans.

◀ *A giant centipede coils its body around a mouse to subdue it. The centipede injects its prey with venom from its maxillipeds.*

COMPARATIVE ANATOMY

Amazing defenses

Millipedes do not have the luxury of a venomous bite to deter predators, but some amazing defenses have evolved. Most have repugnatorial glands, which can spray defensive chemicals 3.5 feet (1 m) or more. Polydesmid millipedes also use chemical weaponry against enemies. They store mandelonitrile, a harmless chemical, in their repugnatorial glands. When threatened, the millipede releases mandelonitrile with an enzyme. The enzyme triggers a chemical reaction that converts the mandelonitrile to a combination of extremely toxic hydrogen cyanide gas and benzaldehyde. Benzaldehyde severely irritates the skin, eyes, and lungs of a predator. Perhaps the most remarkable millipede defense is that of the porcupine–Velcro millipede. When attacked by a rampaging ant, the millipede turns and confronts the ant with its rear end. At the end of the millipede's body are brushlike tufts of bristles lined with hooks. The bristles detach easily, and the hooks catch against hairs on the ant's body, immobilizing it. The millipede is now able to escape unharmed; it has enough bristles to fight off many ants and will replace lost bristles at its next molt. For the ant, however, the torment is only just beginning. The more the ant struggles to free itself, the more entangled it becomes. Eventually the ant starves to death.

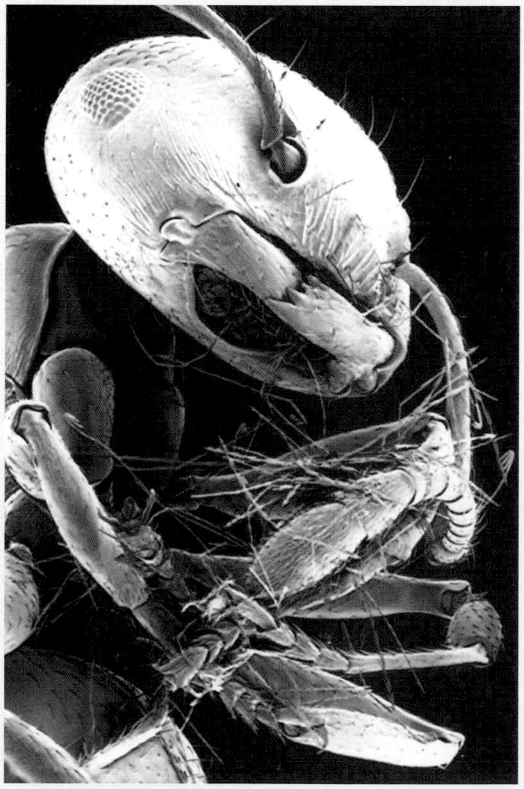

▶ *An ant tries in vain to remove bristles from a porcupine–Velcro millipede. The millipede was able to escape, but the ant died.*

Reproductive system

Female centipedes have a single, large ovary, inside which eggs are produced. A tube called the oviduct runs from the ovary; a structure called the seminal receptacle connects to it, as do several glands. The oviduct links the ovary to a gonopore, or exit from the body, that lies on a legless segment near the tip of the body. A pair of grasping structures called gonopods line the gonopore. Male centipedes have between one and 26 sperm-producing testes that lie above the gut. Sperm is parceled into packages called spermatophores.

Since centipedes are killers that use deadly venom, many species have evolved elaborate courtship rituals in which they "dance" around each other in a circle and stroke each other with their antennae. These displays help each centipede avoid being mistaken for potential prey by the other. The male deposits a spermatophore onto a net woven from silken threads. Then the female grasps the spermatophore with her gonopods and inserts it into her gonopore. The spermatophore remains in the seminal receptacle for a time before fertilization takes place.

Some female centipedes lay a single fertilized egg into a hole in the soil, then abandon it. Secretions from glands along the oviduct coat the eggs. The secretions contain chemicals that kill fungi, which can destroy the eggs. When the young hatch they have seven pairs of legs, but they gain an extra segment and pair of legs each time they molt their outer skin, or exoskeleton. Other female centipedes are more dedicated mothers. They dig small chambers into which they lay small egg clusters. These centipedes guard their eggs fiercely. After hatching, the adult centipede stays with her young until they have molted a few times and are able to hunt for food. Young of these types of centipedes have the same numbers of legs and segments as the adults.

JAMES MARTIN

FURTHER READING AND RESEARCH

Ruppert, Edward E., and Robert D. Barnes. 1994. *Invertebrate Zoology*. Saunders College: Fort Worth.

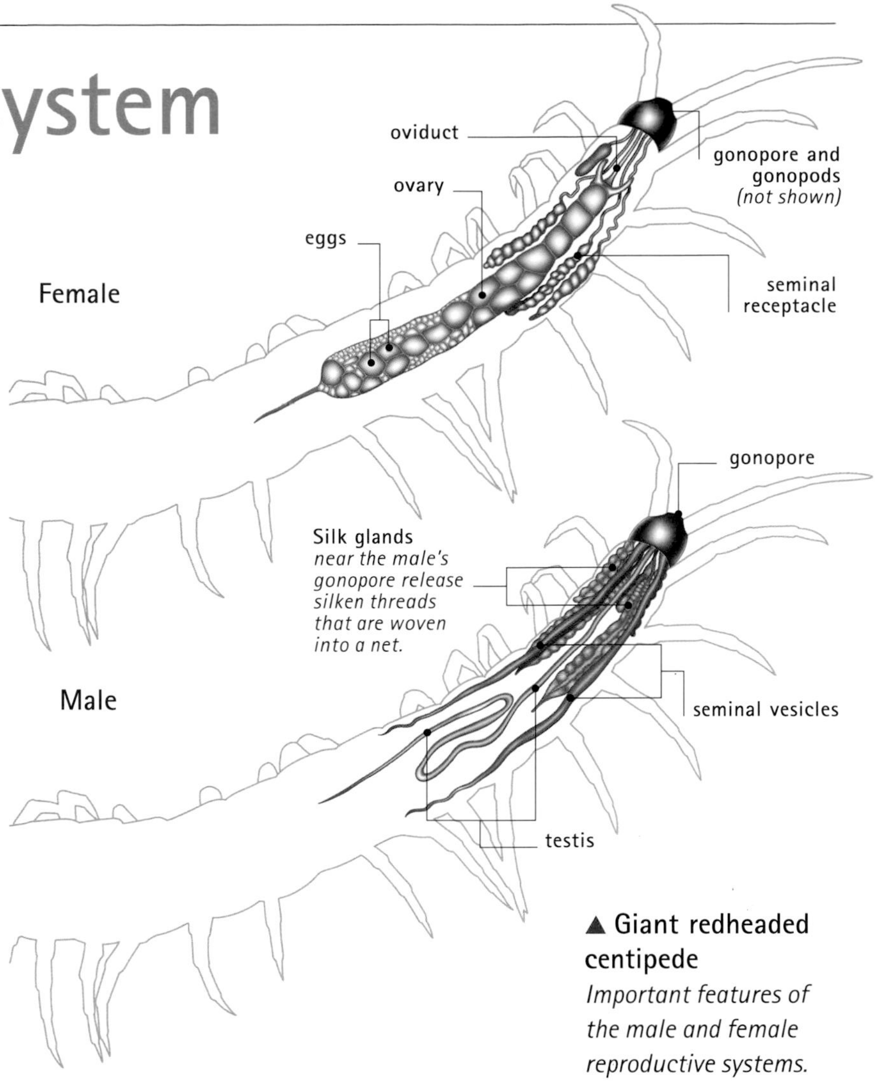

▲ **Giant redheaded centipede**
Important features of the male and female reproductive systems.

COMPARATIVE ANATOMY

Sperm stalks

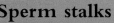

Symphylans are tiny relatives of centipedes with an unusual way of getting sperm to eggs. Males attach their spermatophores to long, flexible stalks that they secrete from glands around their gonopores. When a female symphylan chances upon one of these structures, she bends it using her antennae before biting off the spermatophore. She then stores the spermatophore in her mouth. When her eggs are ready for fertilization they move along the oviduct to the female's gonopore. She collects the eggs with her mouthparts. Coating each with stored sperm from her mouth, the female symphylan then cements her eggs to particles of soil or to moss, where they develop.

Giant clam

ORDER: Veneroida FAMILY: Tridacnidae
GENERA: *Hippopus* and *Tridacna*

Giant clams live in clear, shallow tropical waters, typically on coral reefs. They are the largest bivalve mollusks. Like corals, giant clams get most of their food from microscopic algae called zooxanthellae, which live in their soft tissues. Zooxanthellae use sunlight to make carbohydrate food by a process called photosynthesis. Clams absorb some of this food, while providing the zooxanthellae with nutrients.

Anatomy and taxonomy

Organisms are classified in groups that form part of larger groupings. The classification is based mainly on shared anatomical features, which usually indicate that the members of a group have the same ancestry. The classification shows how the organisms are related to each other. Genetic research is also important for figuring out relationships between organisms. Scientists can also compare the anatomy of living organisms and fossil forms preserved in rocks of known age. That can reveal how long the various groups have existed, and which evolved first.

• **Animals** Animals are multicellular (many-celled) organisms with well-developed powers of movement. Unlike plants, algae, and many microorganisms, they cannot use water and simple chemicals in the air, water, or soil to make complex organic food molecules. Animals have to obtain food in ready-made form, typically by eating other organisms. Animals digest the tissues of these organisms, breaking them down into small molecules that are used to provide energy.

• **Mollusks** The mollusks are a phylum of at least 20,000 species of animals that include slugs and snails, chitons, clams, and cephalopods such as squid and octopuses. Mollusks do not have an internal or an external skeleton, but they are often protected by hard calcareous (chalky) shells. A typical mollusk has a head bearing sensory organs, a muscular foot, and a flap of tissue called the mantle that protects the vital organs.

• **Bivalves** The body of a bivalve mollusk is enclosed by a pair of shells, or valves, joined by a hinge. The valves can usually be sealed together to protect the animal from enemies or hostile environmental conditions. A bivalve has no head, but many types have a large muscular foot. The foot is often used for burrowing into the sand or mud. All bivalves are aquatic, and most get their food by filter-feeding particles suspended in the water.

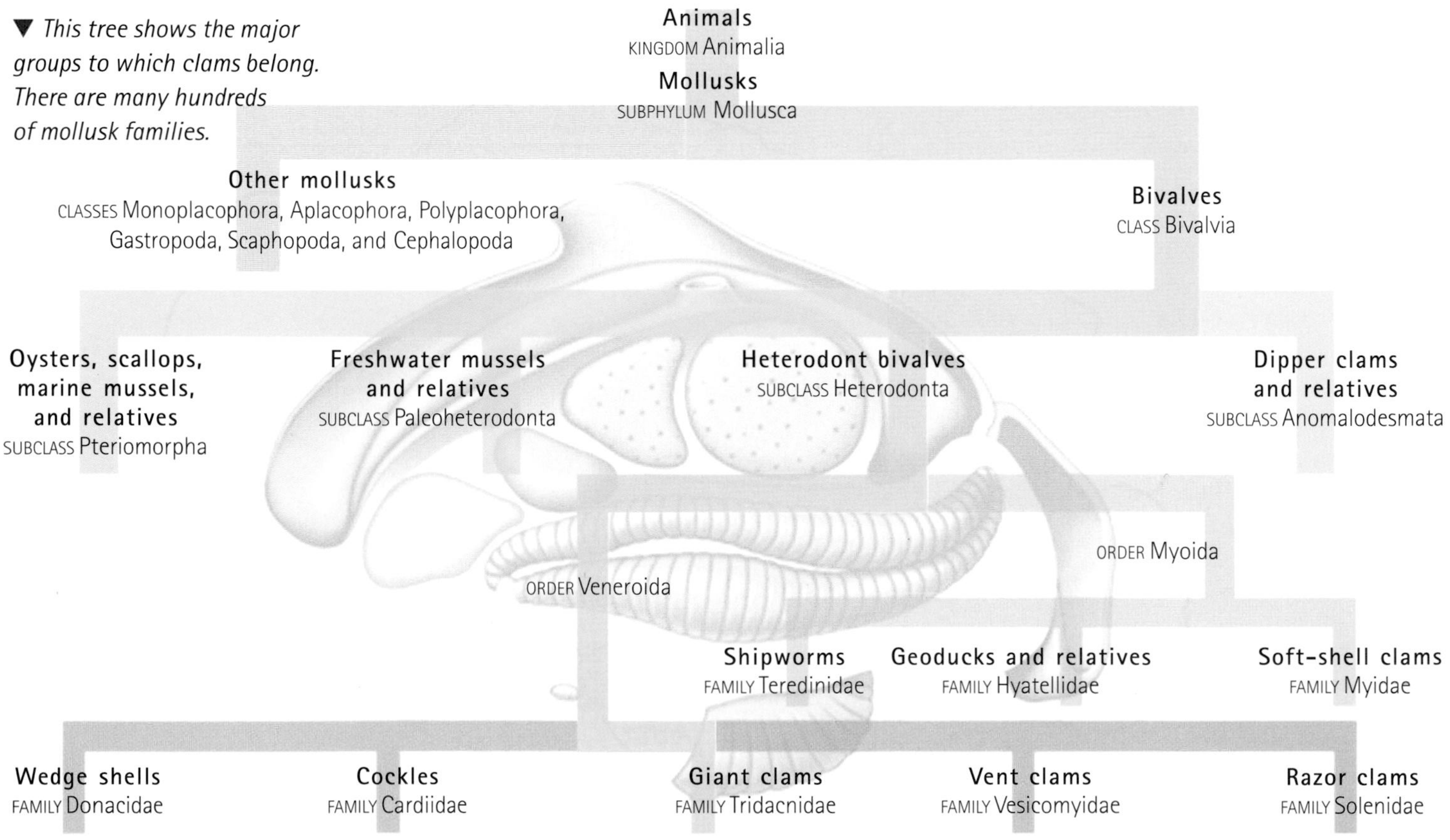

▼ *This tree shows the major groups to which clams belong. There are many hundreds of mollusk families.*

▲ *The giant clam* Tridacna maxima *lives on coral reefs in tropical areas of the Pacific Ocean. This individual, on an Indonesian reef, has animals called tunicates growing on its shell.*

• **Heterodont bivalves** Bivalves that have two types of hinges on their two shells, which are of similar size and shape, form a group called the Heterodonta. Most draw water into and out of their body through tubes called siphons. More than half of the 100 or so families of bivalves belong to this subclass.

• **Veneroida** Veneroida is an order of bivalves that close their shells with a pair of similar-size muscles. This group of bivalves includes cockles, razor clams, vent clams, wedge shells, and giant clams.

• **Giant clams** There are nine species of giant clams in the genus *Tridacna*. These large bivalves root themselves hinge-down in warm, shallow seas. Giant clams draw most of their nourishment from millions of photosynthesizing zooxanthellae that live in their soft mantle tissue.

FEATURED SYSTEMS

EXTERNAL ANATOMY Giant clams can close their tough massive shells for protection. The shells are also often held open, allowing water to move in and out and light to reach algae living inside the mantle. *See pages 344–345.*

INTERNAL ANATOMY Algae live inside a giant clam's mantle; both the clam and the algae benefit from this symbiotic arrangement. The mantle also houses the eyes and protects the delicate internal organs, while the siphons, through which water moves, pass along either side of it. *See pages 346–349.*

MUSCULAR SYSTEM Important bivalve muscle groups include the adductors, which hold the valves together; and the retractors, which many species use for burrowing. *See pages 350–351.*

NERVOUS SYSTEM Bivalves do not have a brain but instead have three pairs of ganglia that coordinate analysis of sensory input and motor responses. *See page 352.*

CIRCULATORY AND RESPIRATORY SYSTEMS A heart pumps blood through vessels. Oxygen uptake occurs through the gills. *See page 353.*

DIGESTIVE AND EXCRETORY SYSTEMS Giant clams get much of their nutrition from symbiotic algae, but they also filter organic particles from the water using their gills and labial palps. *See page 354.*

REPRODUCTIVE SYSTEM Giant clams release sperm and eggs into the water, where they fuse. The young swim for a time as they develop, before settling on coral rubble to grow into adults. *See page 355.*

External anatomy

> **CONNECTIONS**
>
> **COMPARE** the fleshy mantle lips of the giant clam with the feeding tentacles of a ***SEA ANEMONE.*** The lips and tentacles are both soft tissues that do not need much structural support because they are supported by the seawater.

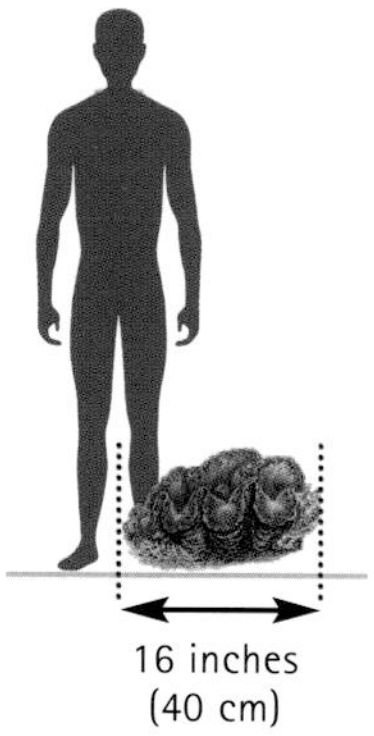

Giant clams can be colossal creatures. The largest ones have a shell length of nearly 5 feet (1.5 m) and weigh more than 2,400 pounds (1,100 kg). Their massive fluted shells look strong enough to withstand the fiercest of blows. Stories have been told of these animals seizing the feet of careless divers and holding them in an unbreakable grip. These stories are myths because a giant clam cannot close its shells tightly enough to grip a human foot. The area between the shells is packed with the clam's soft, colorful mantle tissue, and this stops the shells from closing all the way.

The mantle tissue is the key to a giant clam's survival because it contains colonies of microscopic algae. The algae live in the tissue and use the energy of light to make food. To photosynthesize, the algae must be exposed to bright sunlight, so a giant clam always settles on a hard surface in shallow, clear water with its shell valves opening upward to face the sun. Most other bivalves are burrowing animals that live with their shell valves opening downward.

Strong shells

Most of a giant clam's shell is hidden, because the clam usually settles in a crevice in a coral reef when it is young and gradually fills the crevice as it grows. Eventually the giant clam grows so large that it cannot escape, but that is no problem because the clam has no reason to move. Some species of giant clams, such as the relatively small *Tridacna maxima*, bore into the coral to make a snug home.

▼ VALVES AND MANTLE
Tridacna maxima
This moderate-size species grows to about 16 inches (40 cm) long. It lives in crevices on tropical coral reefs.

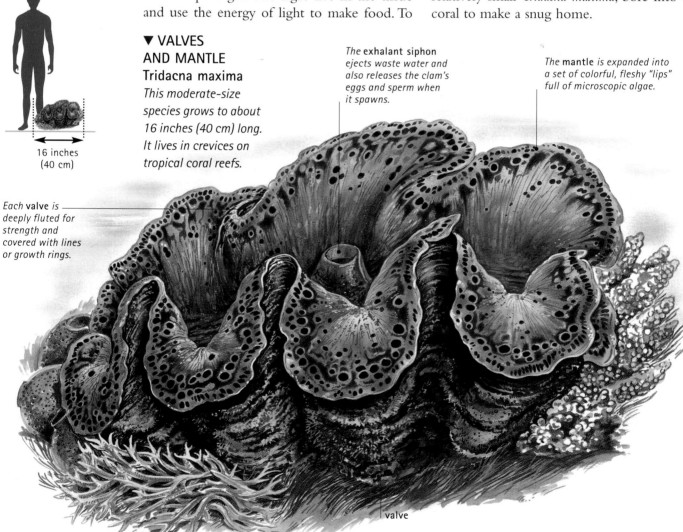

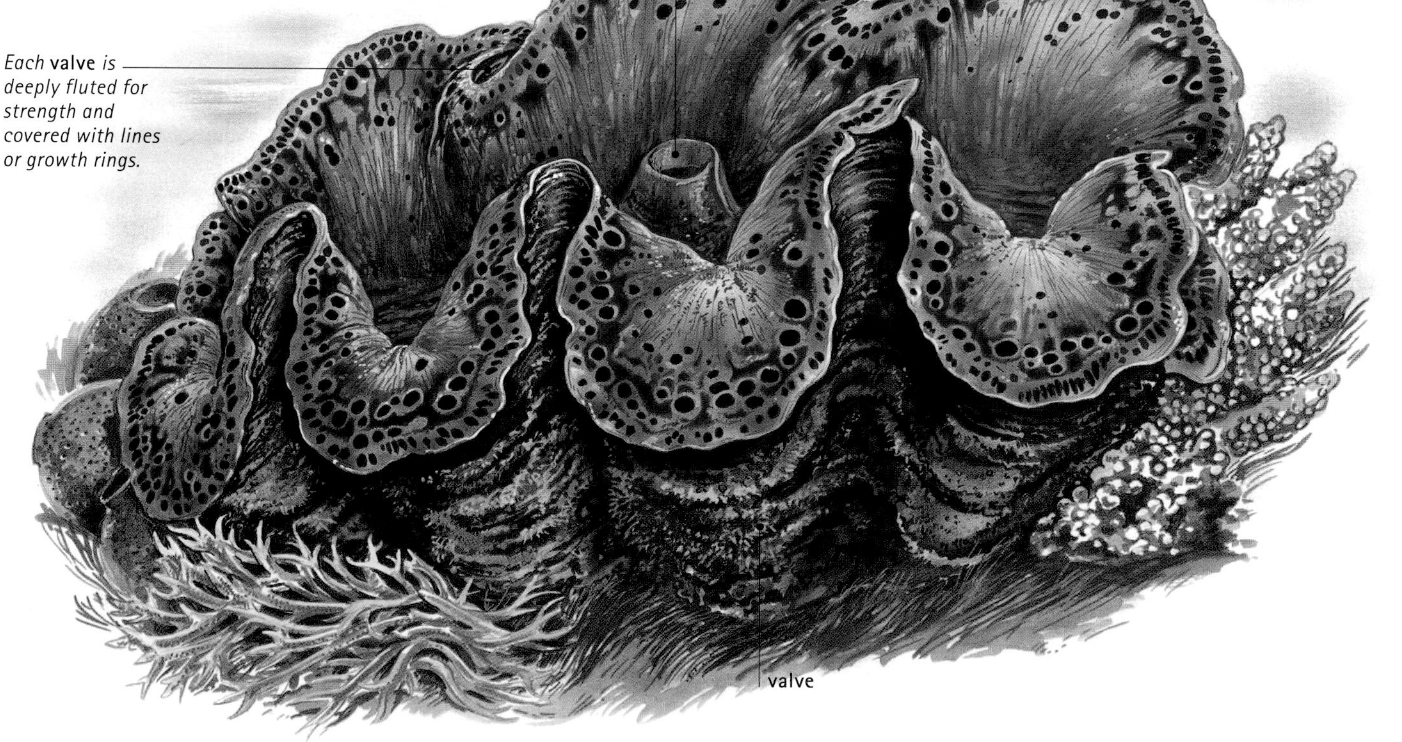

The exhalant siphon ejects waste water and also releases the clam's eggs and sperm when it spawns.
The mantle is expanded into a set of colorful, fleshy "lips" full of microscopic algae.
Each valve is deeply fluted for strength and covered with lines or growth rings.
valve

CLOSE-UP

Steady growth

The material for making a clam's shells, or valves, is produced by cells at the edge of the mantle. The cells add a small amount of new material every day. The main shell material is calcium carbonate, which is the mineral that forms chalk and marble.

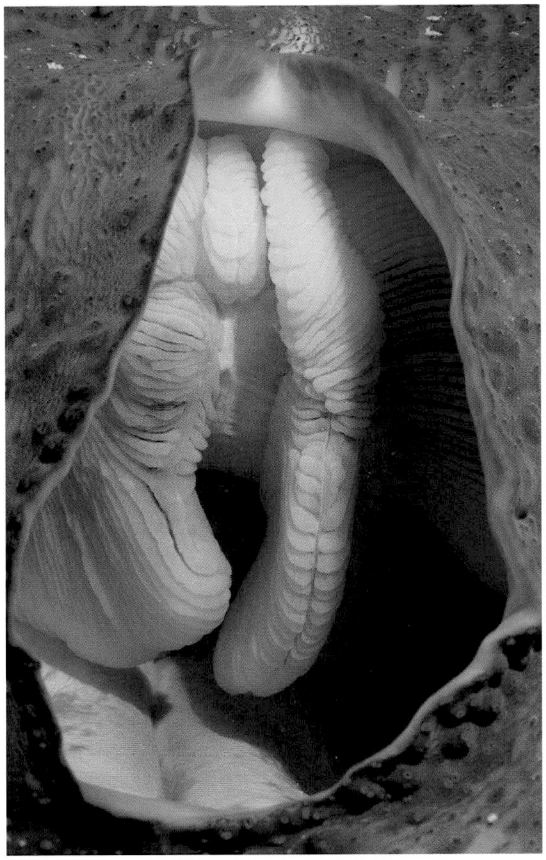

◀ *A close view into the inhalant siphon of the largest species of giant clam,* Tridacna gigas. *The layered structures are gills, through which water is drawn.*

A giant clam's shells, or valves, grow from the hinge region. They grow faster when conditions are good—particularly when conditions are sunny—and more slowly when times are hard. Growth is marked by a series of "growth rings" on each valve. Bivalves such as cockles that live in regions with strongly seasonal climates may stop growing altogether during the winter, leaving very marked growth rings. The number of these rings reveals the cockle's age.

Giant clam valves also have big corrugations, creating large flutes that radiate from the hinge. The flutes make each valve much stronger. Most other bivalves, except the very smallest, have similar ridges, flutes, and ribs. The valves are very alike, so a giant clam is more or less symmetrical. Some bivalves that live on their side, such as scallops and oysters, have one deeply hollowed valve that can hold the animal's whole body, and one flat valve that acts like a lid.

Snorkel tubes

Nestling at the heart of a giant clam's colorful mantle lips is a round aperture-like a funnel. It is the clam's exhalant siphon—an outlet for the water that the animal sucks in to obtain oxygen and some of its food. A giant clam also has an inhalant siphon through which water enters the mantle cavity.

The siphons of a giant clam are short, but some burrowing bivalves have long tubes that allow them to draw water from above the mud when they are buried deep below the surface, or substrate. The most impressive burrowing bivalve is the geoduck, which lives along the Pacific coast of North America. A geoduck's siphon is so long that the bivalve can live more than 39 inches (1 m) beneath the seabed.

COMPARATIVE ANATOMY

High-speed mollusks

They seem like completely different types of animals, but clams and squid are both mollusks. They belong to different classes (clams are bivalves, while squid are cephalopods) but they both have a soft body, a mantle cavity, and siphons. A squid draws water into the mantle cavity and then squirts it out through the siphon. This jet of water propels the animal swiftly through the water. Squid are high-speed hunters with a relatively large brain and acute vision, quite unlike the giant clam, which is immobile and has no head.

Maneuverable jets

Squid can move by rapidly ejecting water from their mantle, and they change direction by altering the angle of the siphon.

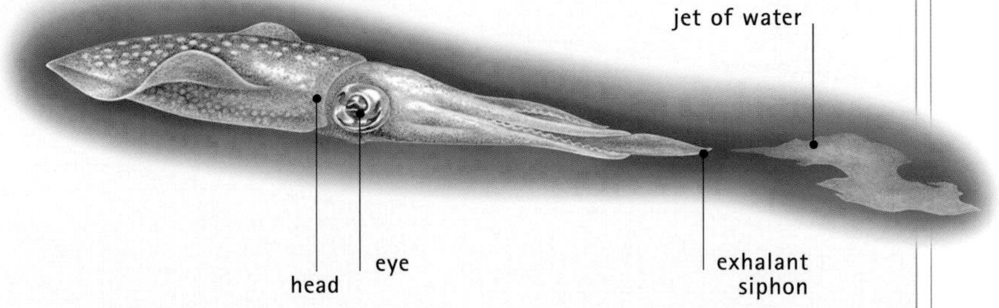

Internal anatomy

CONNECTIONS

COMPARE the gill arrangement of a giant clam with the system in the ***SAILFISH***. Both animals get their oxygen from seawater but in different ways. A giant clam has to pump water through its siphons, but a fast-swimming sailfish can depend on water being forced through its gills as it swims.

Bivalves, such as giant clams, do not have a head, and there is no clear pattern to the arrangement of the body and internal organs. Despite these factors, the giant clam is a complex animal. Other headless animals such as starfish, sea anemones, and jellyfish are radially symmetrical, with organs and appendages radiating from a central hub. Although these animals have no "front end," they do have a center. However, the internal anatomy of a bivalve is not radially symmetrical. A bivalve's body contains a collection of organs packed into an enclosing pair of shells.

Mantle and siphons

The most conspicuous part of a giant clam is its colorful mantle. This flesh is an extension of the flap of soft tissue that encloses the delicate gills and other internal organs of all mollusks. In bivalves, the mantle lines each shell and is attached along the pallial line, a short distance from the edge of the shell. The pallial line is clearly visible on empty bivalve shells.

The mantle tissue lining each shell of a bivalve may be separate, but usually the two sides are at least partially fused. This arrangement helps keep sand and mud out of the body of burrowing bivalves such as cockles, razor clams, and wedge shells. The mantle of a giant clam is fused over most of its length, increasing the surface area that can be colonized by algae and exposed to sunlight.

A giant clam has to draw oxygen-rich water into its body and pump it out again, so there are always gaps in the fused mantle to allow the water in and out. In most bivalves the gaps are formed into tubes or siphons, but the inhalant siphon of a giant clam is just a simple gap in

▶ *This diagram shows important parts of a giant clam's various internal systems, including the muscular, digestive, respiratory, and circulatory systems.*

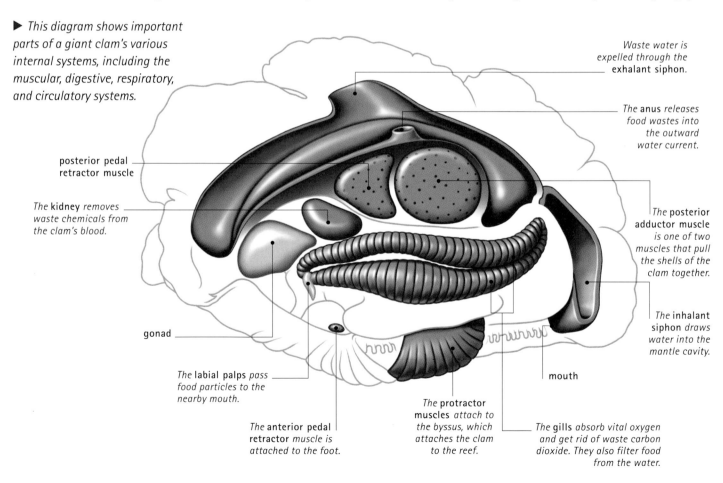

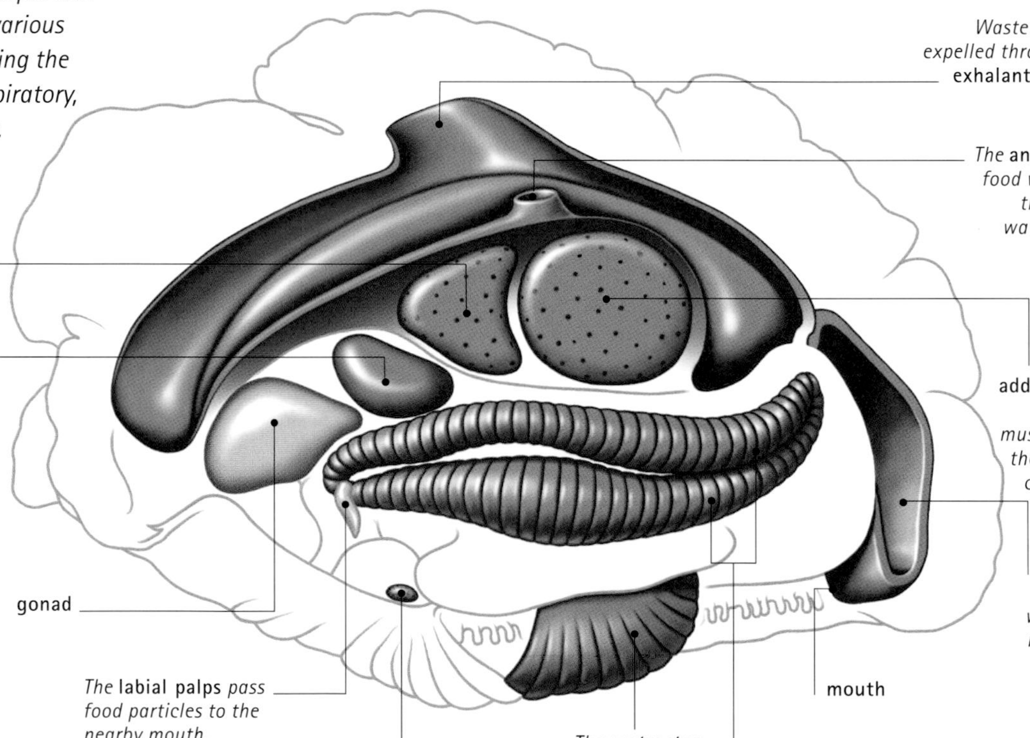

Waste water is expelled through the exhalant siphon.
The anus releases food wastes into the outward water current.
posterior pedal retractor muscle
The kidney removes waste chemicals from the clam's blood.
The posterior adductor muscle is one of two muscles that pull the shells of the clam together.
The inhalant siphon draws water into the mantle cavity.
gonad
mouth
The labial palps pass food particles to the nearby mouth.
The protractor muscles attach to the byssus, which attaches the clam to the reef.
The anterior pedal retractor muscle is attached to the foot.
The gills absorb vital oxygen and get rid of waste carbon dioxide. They also filter food from the water.

the mantle at the side. The exhalant siphon is more tubelike and is formed from an extension of the mantle tissue. However, the exhalant siphon is very short compared with the siphons of deep-burrowing species like soft-shell clams and the geoduck.

Gills and water flow

Water that flows into the clam's body through the inhalant siphon passes over and through feathery gills, which fill a large part of the body cavity. Oxygen moves from the water through the gills, and waste carbon dioxide passes the other way. The water then passes out through the exhalant siphon.

All bivalves use this system of gaseous exchange. Gills usually serve some other functions as well. Many bivalves use gills to gather small particles of food from the water. A giant clam does this, too, even though much of its food is made by the algae living in its mantle. Food particles collected on the gills are passed to structures called labial palps, which sort the particles and pass them to the clam's mouth. Anything inedible is forced out of the inhalant siphon whenever the clam squeezes its shells together. Any waste left after digestion passes out of the anus and is carried out of the exhalant siphon with the waste water.

IN FOCUS

Resident crabs

The huge mantle cavity of the giant clam *Tridacna gigas* makes a perfect home for creatures called pea crabs. As their name suggests, most pea crabs are pea size, but the species that makes its home within the mantle of a giant clam is much bigger. Even so, it is still only 0.75 inch (20 mm) across its shell. Despite all the space in a giant clam there are never more than two crabs in each one—one male and one female. These two get rid of trespassing crabs by killing and eating them.

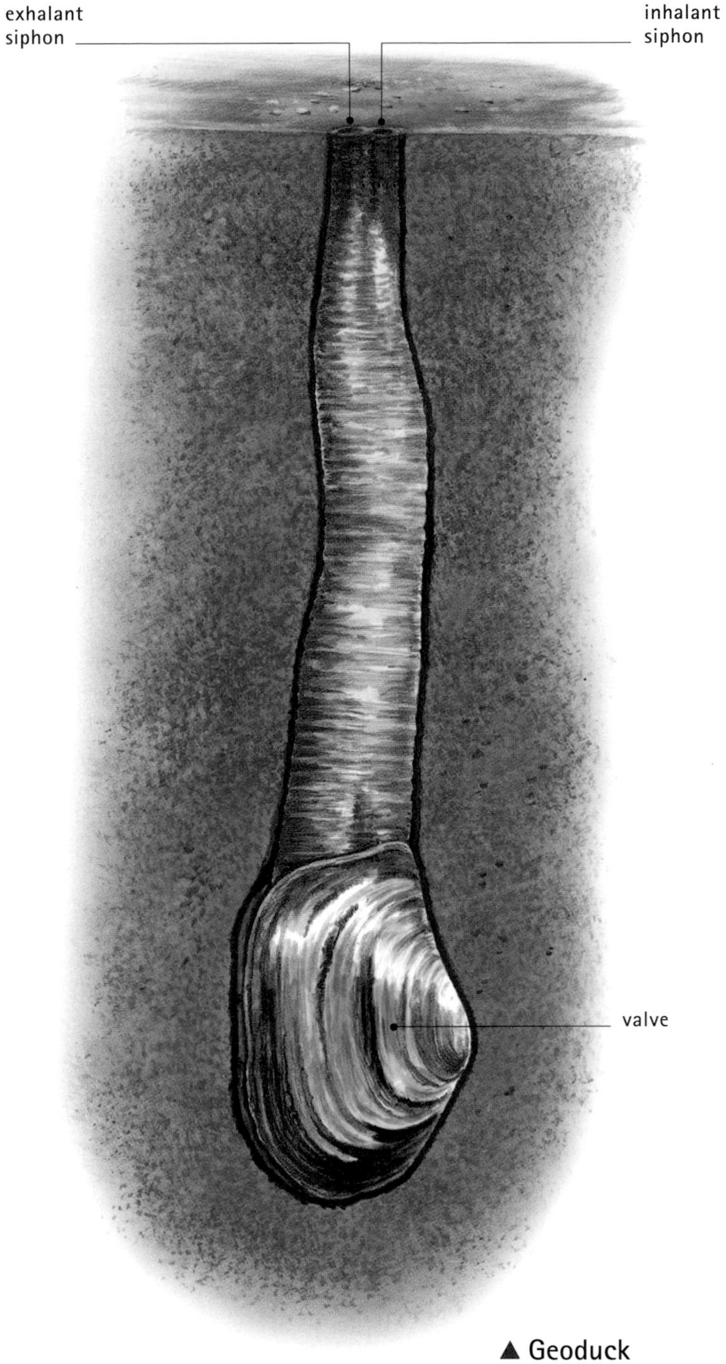

▲ Geoduck
The geoduck (pronounced gooey-duck) lives buried in the sand. It breathes by extending its long siphon to the surface of the sand.

Wired down

A giant clam's shells are pulled together by powerful adductor muscles that pass from one valve to the other. Some bivalves such as oysters and scallops have just one central adductor muscle, but most—including giant clams—have two. A giant clam also has two retractor muscles that are attached to the animal's single foot. The foot of a typical burrowing bivalve such as a cockle or wedge shell is a lobe of muscular tissue that can be extended to get a grip in sand or mud. The foot of a giant clam is different. Early in life the clam uses its foot to anchor its body to the substrate. A gland in the foot produces threads (called byssuses) of a tough, gluelike protein. These keep the clam from being swept away by currents flowing over the reef.

Some of the larger species of giant clams lose their byssal threads as they grow. Instead, they rely on their immense weight to hold them securely in place in their crevice. Smaller bivalves that anchor themselves to rocks rely on their byssal threads throughout their lives. These bivalves include the common mussels that attach to rocks and wharves on tidal shores. Each byssal thread is glued to the rock by the mussel's fingerlike foot. The byssal threads of some types of mussels resemble metallic wire; in medieval times people gathered the threads and wove them together to make "cloth of gold."

IN FOCUS

Precious pearls

The lining of a clam's shell is made of flat calcium carbonate crystals that form a substance called nacre. In many bivalves the nacre gleams with pearly iridescence and is often called mother-of-pearl. That is a good name because bivalves such as oysters and freshwater mussels use exactly the same substance to make pearls. The process starts when a small mineral grain becomes lodged in the mollusk's mantle. The mollusk smothers the irritating particle by releasing layers of nacre from the mantle. In time, this secretion hardens into a pearl.

▼ *The shiny, round object in the mantle of this oyster is a pearl. Oysters, clams, and mussels are all bivalve mollusks.*

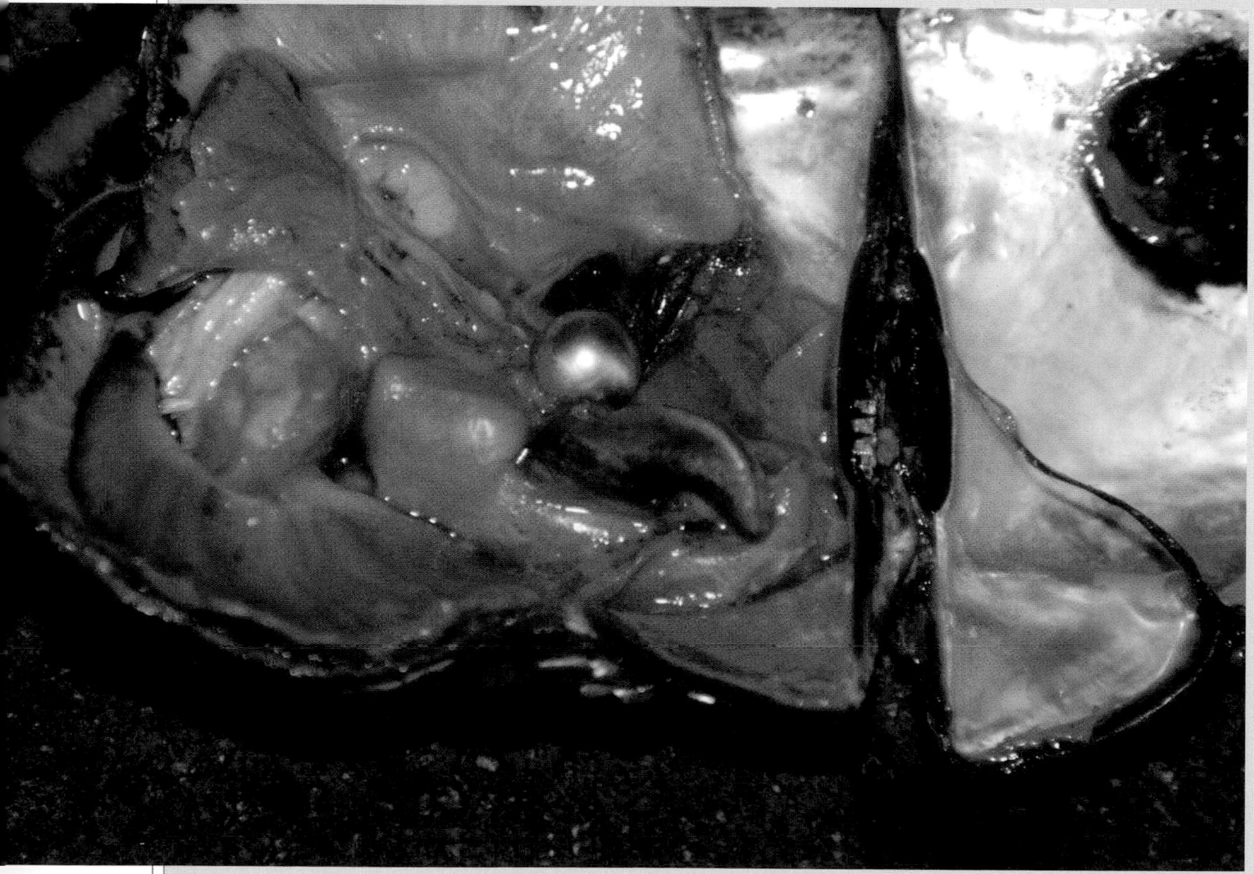

▼ *Oysters attach themselves to the seafloor with glue that they produce inside their mantle.*

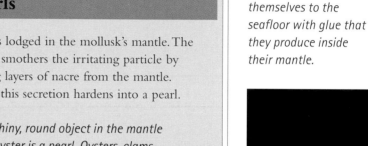

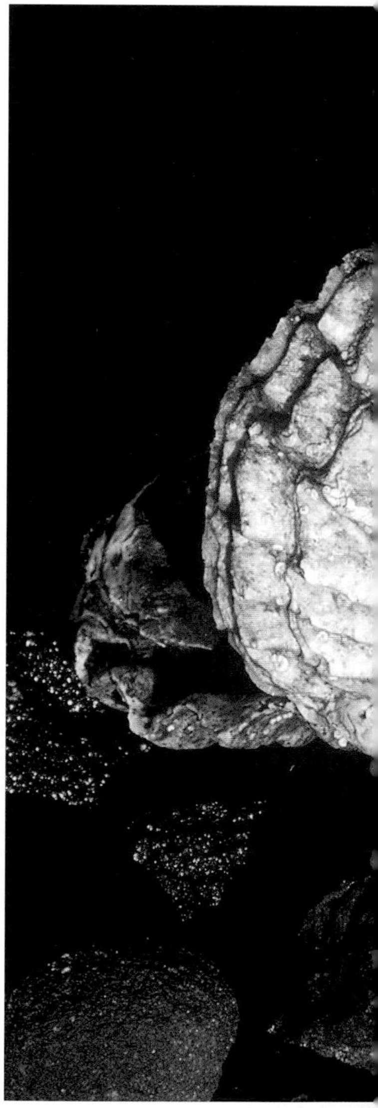

Spring-loaded shells

The foot of a giant clam lies next to the shell hinge, which has interlocking "teeth" that keep the shell valves in line. The hinge is held together by two ligaments of flexible, elastic protein. When a giant clam closes its shells by contracting its adductor muscles, the outer ligament is stretched while the inner one is squeezed. When the clam relaxes its adductor muscles, the springiness of the ligaments makes the shells open again. All bivalves have a similar system of muscles and ligaments for opening and closing their shells. When a bivalve dies, its adductor muscles can no longer hold the shell closed against the spring of the elastic ligaments. That is why the empty shells of bivalves that litter many beaches are always open.

COMPARATIVE ANATOMY

Glued-down oysters

Giant clams and mussels use byssal threads to attach themselves to rocks and corals. However, many oysters, including familiar species like the American oyster, use a different system to secure themselves to the ocean floor. They produce a small amount of natural adhesive (gluelike substance) from the mantle. The young oyster positions itself upon the glue so that the larger of its two valves becomes stuck to the hard surface. Once its valve is cemented in place an oyster cannot move, except to open its upper free valve to draw in water.

Muscular system

CONNECTIONS

COMPARE the muscle system of a sessile (attached, or stationary) mollusk such as a giant clam with the more complex muscle system of a mobile mollusk such as an ***OCTOPUS***.

A giant clam does not need to work hard for a living. It simply spreads its vividly colored mantle in the warm, clear, sunlit seawater and lets the algae in its tissues produce its food. Compared with most bivalve mollusks, a giant clam makes little use of its muscles.

The most active muscles are the pallials of the inner mantle. The mantle is the soft, fleshy outer tissues of the clam's body; it is adjacent to the inside of the valves. The pallials pump oxygen-rich water through the clam's mantle cavity. The clam cannot survive without oxygen, so the pallial muscles of giant clams and other bivalves are always working, just like the muscles that people use for breathing.

A giant clam also uses its adductor muscles frequently. The adductors close a bivalve's shells. The muscles pass from one shell valve to the other through the bivalve's body. Their anchor points are marked by scars on the shell lining. These marks can be seen on any empty bivalve shell. The adductor muscles work against the spring of the hinge ligaments, so if a clam needs to pull its valves together for a long time, it must keep its adductor muscles contracted.

The ability to contract the adductors for long periods is vital for bivalves that live on tidal shores and do not bury themselves in wet sand or mud. When the tide ebbs, for example, a mussel contracts its adductor muscle to keep from drying out. The water stored inside its shell also contains the oxygen it needs and helps it keep cool in the summer sun. The mussel has to use its adductor muscles to hold its shell tightly shut for up to six hours until the tide covers the animal again. Many burrowing mollusks such as cockles do the same, but since they are less likely to dry out they do not have to stay closed for so long.

Bigfoot bivalves

Apart from the pallials and adductors, a giant clam's other main muscles are the two retractor muscles; there is a big one in the middle of the

▶ **MUSCLE CROSS SECTION**
There are four main sets of muscles in a giant clam's body: the adductors, retractors, protractors, and pallials. The last, which connect the shell and the mantle, are omitted for clarity.

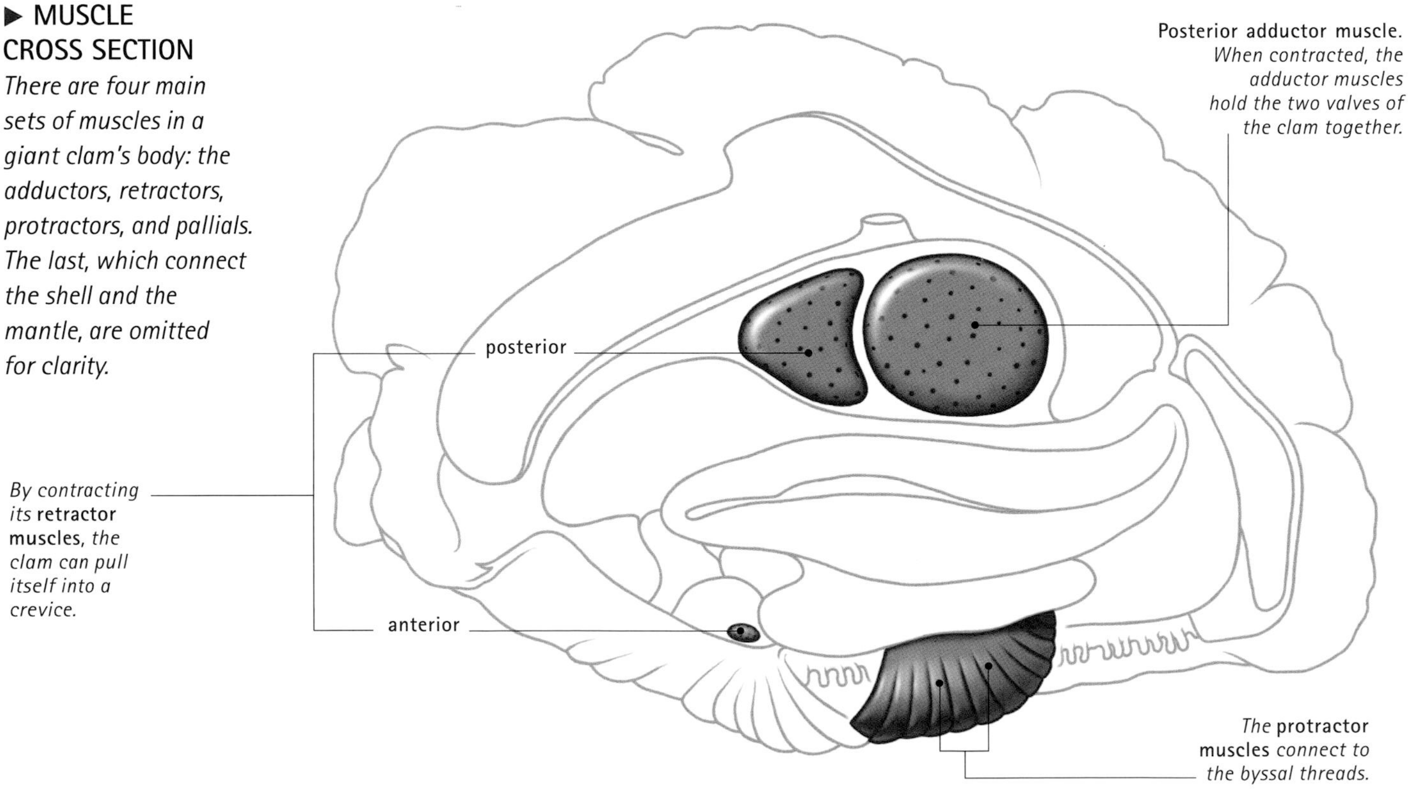
Posterior adductor muscle. When contracted, the adductor muscles hold the two valves of the clam together.
posterior
By contracting its retractor muscles, the clam can pull itself into a crevice.
anterior
The protractor muscles connect to the byssal threads.

animal, and a smaller one near the shell hinge. Retractors work the clam's foot, which is attached to the coral below. By contracting its retractor muscles, the clam can pull its body down into a crevice. This may be useful if the tide is flowing fast across the reef, especially for young, lightweight clams. The biggest giant clams do not attach to rocks and may never need to use their retractor muscles.

Burrowing bivalves use their retractors all the time. A razor clam, for example, has a large foot that it pushes down into wet sand, using muscles in the foot itself called protractor muscles. When the foot is pushed down as far as it will go, a razor clam pumps blood into it, which makes it expand into an anchor. The clam then contracts two pairs of retractor muscles, which pull the clam down into the sand. The protractor and retractor muscles work fast, and they are also very strong.

Wedge shells live on surf beaches and are amazing burrowers. A wedge shell can move so rapidly using its retractor muscles that it can back out of its burrow to catch a passing wave, let the wave carry it up the beach, then quickly bury itself again as the wave falls back. A cockle can even use its foot muscles to jump through the water and escape its enemies.

IN FOCUS

Muscle fibers

The muscles of most bivalves are made of two types of fibers (long, thin cells): smooth and striated. Striated fibers work fast, triggered by commands from the nervous system. When a bivalve such as a mussel needs to shut its shell suddenly, striated fibers in its adductor muscles do the job. Although these fibers work fast they cannot stay contracted for long periods, such as when the tide goes out. That is the job of smooth muscle fibers, which work slowly but are able to keep the shell closed for hours.

Striated muscle fibers

Smooth muscle fibers

PREDATOR AND PREY

Jet propulsion

Scallops are bivalves that live on the surface of the seabed. They have an unusual way of escaping enemies such as starfish. As soon as a scallop senses that a starfish is nearby, it claps its valves together. This action pumps water from the mantle cavity through small gaps between the valves. The water squirts out of the gaps in powerful jets. This movement sends the scallop shooting through the water for 3 feet (1 m) or more.

1. A scallop lies on the seabed, with its long, sticky feeding tentacles exposed to gather food.

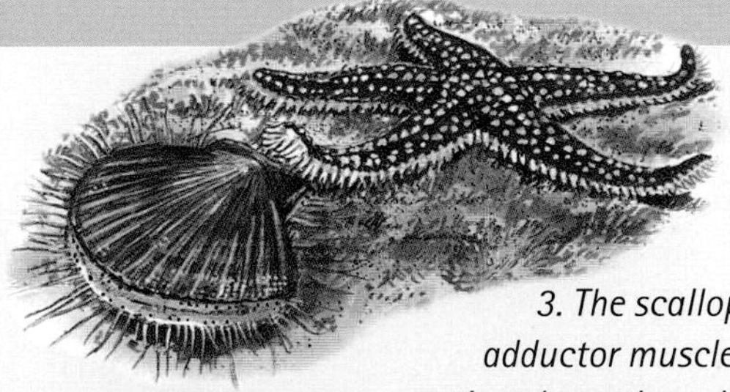

2. The scallop detects the touch of a predatory starfish.

3. The scallop contracts its adductor muscles quickly. This action closes the valves and forces out a jet of water that propels the mollusk away from danger.

Nervous system

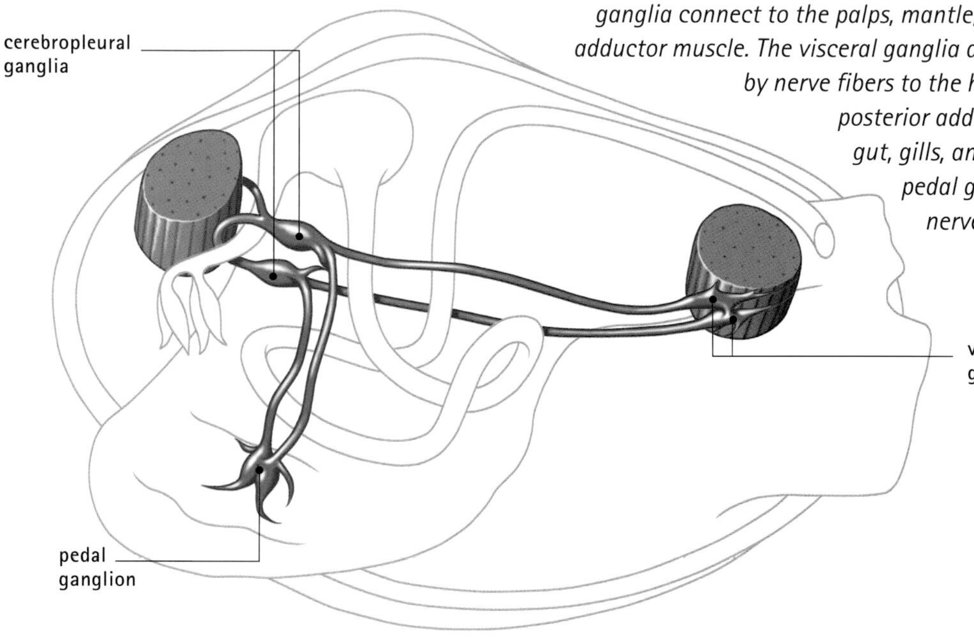

◀ Bivalve
This illustration shows nerves and ganglia of a typical bivalve. Nerve fibers from the cerebropleural ganglia connect to the palps, mantle, and anterior adductor muscle. The visceral ganglia are connected by nerve fibers to the heart, mantle, posterior adductor muscle, gut, gills, and siphon. The pedal ganglia link to nerves in the foot.

CONNECTIONS

COMPARE the nervous system of the giant clam with the nerve net of a ***JELLYFISH***. The nerves of a jellyfish are not focused on ganglia like those of a clam. All the actions of a jellyfish are automatic. The ganglia of a giant clam might allow the animal to make simple choices.

It seems unlikely that an animal without a head, such as a giant clam or another bivalve, can have a complex nervous system. Yet clams react swiftly to things that happen around them. Dipper clams can detect the movements of live prey and use their siphon tubes to catch it. A scallop can see the shadow of an approaching enemy, using the many small eyes at the edge of its mantle. It can also feel and taste things with its fringe of sensory tentacles. Other bivalves have nerves on their siphon tubes that check the flow, temperature, and saltiness of water. Keeping track of such things is vital for bivalves that live on tidal shores.

IN FOCUS

Bivalve eyes

The colorful mantle of a giant clam contains several thousand simple eyes. Each eye is a hollow cup lined with light-sensitive cells. When light enters the cup, some cells are lit up while others remain in shadow. This lets the clam know the source of the light. The cup can also detect nightfall and the shadows of possible enemies, but a clam cannot see images in the way that humans can. It is possible, however, that a scallop can see some sort of image, because each of the many beadlike eyes around its mantle rim has a lens to focus the light. These lenses make the scallop's eyes far more efficient than the eyes of a giant clam.

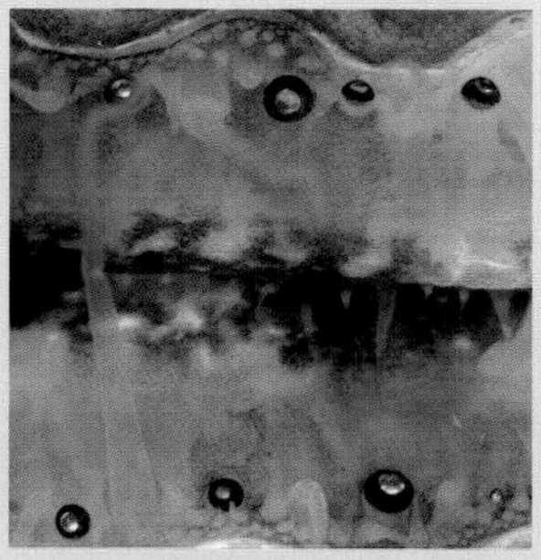

SCALLOP EYE
The tiny eyes of a scallop line the edge of the mantle and enable the scallop to see simple images.

Nerve centers

All this information is gathered by sensory cells. They send electrical signals that pass down nerve fibers into two pairs of long nerve cords. These are linked to three pairs of nerve centers called ganglia. The pedal ganglia are present in the bivalve foot and connect via nerve cords to the cerebropleural ganglia. These, in turn, connect to the visceral ganglia next to the posterior adductor muscle. The two cerebropleural ganglia are also connected to each other by a dorsal commissure, which loops over the esophagus.

Ganglia process simple information, but they cannot do the complex tasks handled by a true brain. However, they do control the animal's muscles and digestive system, help it respond to its environment, and keep it out of danger.

Circulatory and respiratory systems

Like all animals, the giant clam needs a system for gathering oxygen and getting rid of the carbon dioxide produced by its bodily processes. Terrestrial animals breathe air, which contains oxygen, but the clam must absorb oxygen that has been dissolved in water. It does this using a pair of gills.

Each gill is a featherlike series of tubes with very thin walls. Blood flows through the tubes, which are surrounded by water drawn in through the inhalant siphon. Oxygen passes from the water, through the thin gill walls, and into the blood. Carbon dioxide goes in the opposite direction—out of the blood and into the water. The water then flows out of the clam's exhalant siphon.

Some bivalves, including the giant clam, are also able to exchange gases through their mantle tissue. Some clams that are exposed to the atmosphere at low tide are even able to obtain oxygen directly from the air.

Heart and arteries

Clams have a simple heart. The heart pumps oxygen-carrying blood through a system of arteries to the muscles and internal organs. The blood delivers oxygen and sugars (made by algae that live in some of the clam's tissues) that fuel the clam's body processes. The blood also collects carbon dioxide and other waste materials. This waste-carrying blood flows through the kidney, or nephridium, which removes unwanted chemicals. The blood then returns to the gills, where it loses its carbon dioxide and picks up more oxygen.

COMPARATIVE ANATOMY

Doing without blood

A clam uses circulating blood to carry oxygen absorbed by its gills to every part of its body. Some other organisms, such as insects, do not use the blood for oxygen transport. Their blood instead transports nutrients and wastes. Jellyfish and sea anemones live without any blood. Their body is made of just two layers of functioning cells, divided by a thick layer of gel-like substance called mesoglea. All the cells are in contact with the water. So each cell can absorb all the oxygen it needs directly from the water and lose carbon dioxide in the same way.

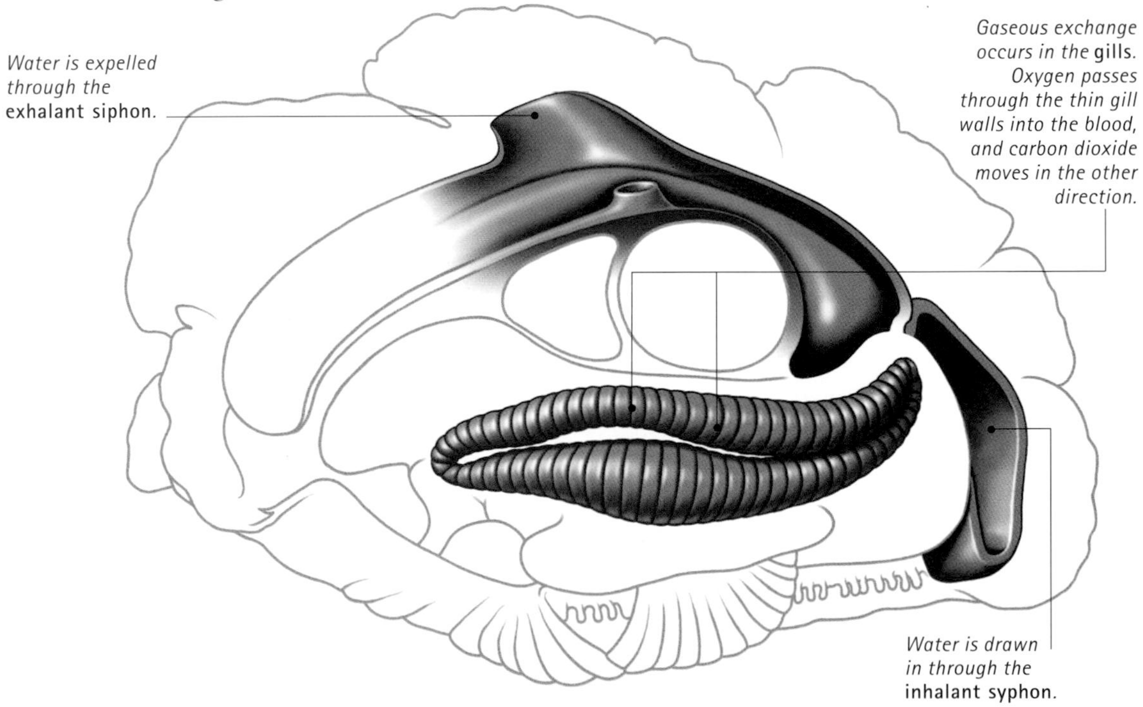

Water is expelled through the exhalant siphon.
Gaseous exchange occurs in the gills. Oxygen passes through the thin gill walls into the blood, and carbon dioxide moves in the other direction.
Water is drawn in through the inhalant syphon.

▶ RESPIRATORY SYSTEM

The giant clam draws water into its inhalant siphon and over its gills, There, oxygen passes from the water through the thin walls of the gills. The oxygen passes into a series of thin tubes that carry blood around the clam's body.

Digestive and excretory systems

Giant clams have most of their food made for them by the microscopic algae, or zooxanthellae, that live in colonies in their mantle folds. However, the clams also gather food particles from the water they pump through their gills. This filter-feeding system is typical of many kinds of bivalves.

The delicate gills of a mollusk are protected by sticky mucus. Simple bivalves like nut clams use the mucus to remove any grit or other debris that clogs their gills. More complex bivalves such as giant clams also gather food particles in the mucus and pass it along the gills to their labial palps and mouth. In the stomach the strings of mucus and food are mixed with digestive enzymes (chemicals that break down food) produced by a structure called the crystalline style. The mixture then passes to small sacs called digestive diverticula, where the digested food is absorbed.

Some bivalves do not have a digestive system. Vent clams live around hot vents in the deep ocean. They simply absorb food made by colonies of bacteria living in their gills. The bacteria make the food using energy from sulfide chemicals that the clam gathers from the hot vent. The awning clam uses the same system, but gets its chemicals from sludge produced by wood pulp mills in the Pacific states of the United States.

IN FOCUS

Perfect partners

The zooxanthellae that live in the mantle of a giant clam are microscopic single-celled organisms. They are able to make sugar and other carbohydrates from water and dissolved carbon dioxide, using the energy of sunlight. This process is called photosynthesis. Zooxanthellae also need other substances, including nitrates and phosphates, which they use to make proteins. Proteins are the main building blocks of living things. The zooxanthellae get these other substances from the giant clam, which obtains them from the food that it gathers by filter-feeding. So although the clam needs the zooxanthellae to supply it with sugar, the zooxanthellae need the clam to supply them with nutrients. This kind of relationship, in which both partners benefit, is called mutualism.

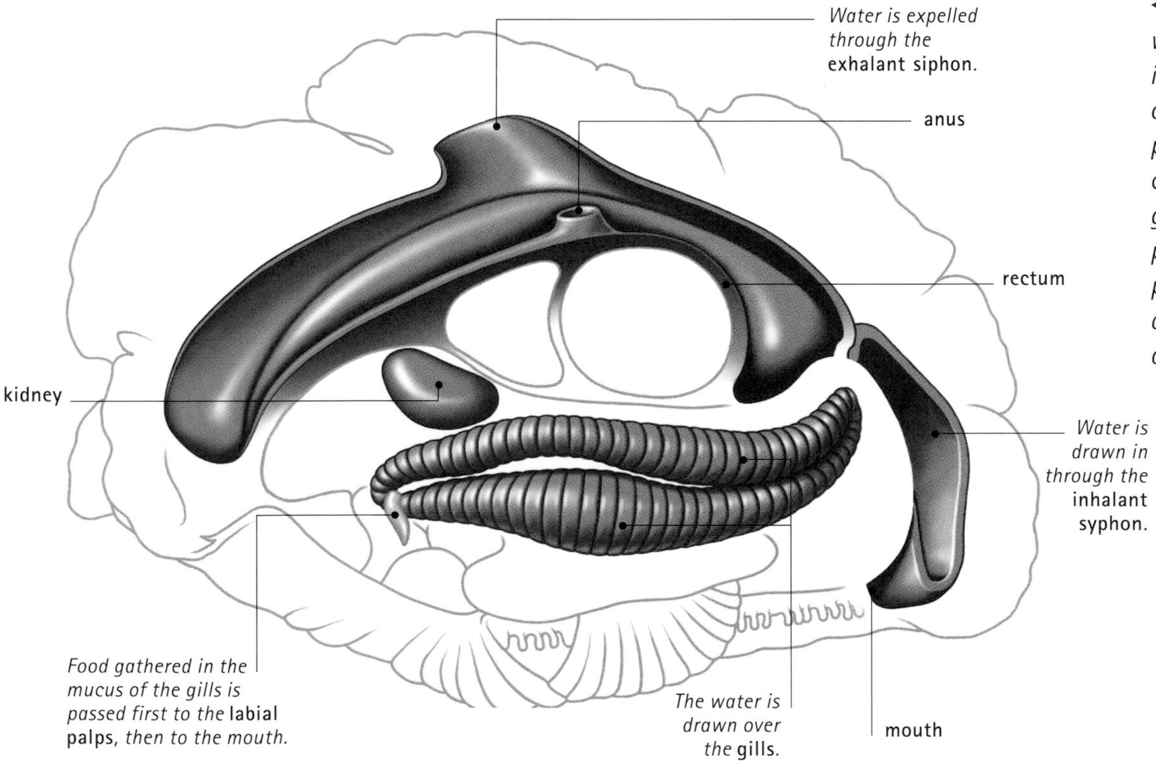
Water is expelled through the exhalant siphon.
anus
rectum
kidney
Water is drawn in through the inhalant syphon.
Food gathered in the mucus of the gills is passed first to the labial palps, then to the mouth.
The water is drawn over the gills.
mouth

◀ *Giant clams draw water into their inhalant siphon and over the gills. Food particles in the water collect in mucus in the gills. The food then passes to the labial palps, into the mouth, and through the digestive system.*

Reproductive system

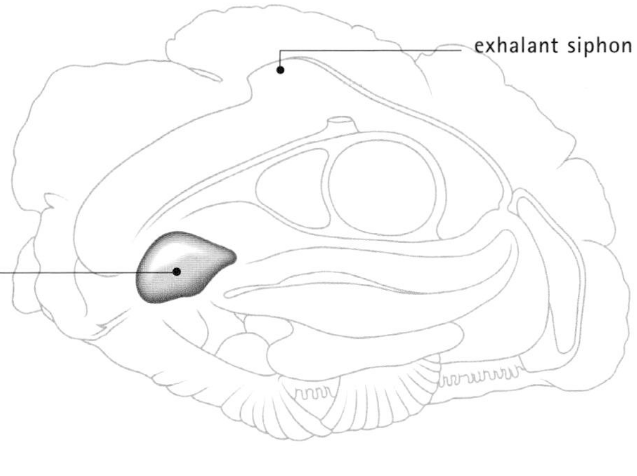

▶ MALE SEX ORGANS
Sperm from the giant clam's two gonads pass into the mantle cavity, then out through the exhalant siphon into the ocean water. There, they may fertilize eggs released by other individuals. Each fertilized egg hatches into a larval clam.

Rooted in its crevice, a giant clam is unable to seek out a breeding partner. Like other bivalves it gets around this problem by pumping millions of eggs or sperm into the water at the same time as other clams. This release of sex cells can be triggered by chemicals produced by other individuals or by the phases of the Moon.

About 12 hours after a giant clam egg has been fertilized by a sperm, it hatches as a tiny trochophore larva. After a day this larva turns into a two-shelled veliger larva that is able to swim and gather food using a structure called the velum. This is covered with tiny hairlike structures called cilia. The veliger feeds on microscopic animals until it is ready to settle in a crevice and develop into an adult clam.

IN FOCUS

Twitching fish lures

Some freshwater bivalves attract fish using animal-like lures. These can be amazingly similar in appearance to the animals they imitate. Part of the mantle of the North American lamp-mussel, a bivalve relative of the giant clam, is shaped like a small fish, complete with fins and eyes. The lamp-mussel twitches the fishlike lure to make it easily noticeable to a predatory fish. When the fish comes close to investigate the lure, the mussel's bloodsucking young use the opportunity to latch onto the fish. Sometimes the lure is swallowed by the fish, and the young break free to attach to the inside of the fish's mouth.

Giant clams are unusual because they start life as males but eventually develop female sex organs as well. Most bivalves are either male or female, although oysters can change sex more than once in their lives.

Freshwater mussels do not scatter eggs and sperm into the water like giant clams. The females draw sperm into their body to fertilize their eggs. The larvae hatch in their gill cavities, and become tiny swimming organisms with a sharp hook on each valve. A larva uses the hooks to latch onto a fish. This allows the larva to grow without being swept downstream. The larva feeds on the fish's blood for up to a month. It then drops to the riverbed, where it develops into an adult mussel.

TREVOR DAY

FURTHER READING AND RESEARCH

Brusca, R. C., and G. J. Brusca. 2003. *Invertebrates.* Sinauer Associates, Inc.: Sunderland, MA.

▼ *Millions of sex cells spout into the ocean from the exhalant siphon of a giant clam.*

Giraffe

ORDER: Artiodactyla FAMILY: Giraffidae GENUS: *Giraffa*

The giraffe is the tallest animal in the world, reaching a height of 18 feet (5.5 m). It lives in savannas and open woodland areas across much of Africa south of the Sahara, but its range has become heavily fragmented over the last hundred years or so. The animal's name derives from *zirafah,* which means "fast walker" in Arabic. Adult giraffes can run at up to 35 miles per hour (56 km/h).

Anatomy and taxonomy

Scientists group all organisms into taxonomic groups based largely on anatomical features. Giraffes belong to the order Artiodactyla, the even-toed ungulates, one of the largest mammal groups. The artiodactyl families closest to the Giraffidae are the Moschidae (musk deer), Antilocapridae (pronghorn antelopes), Cervidae (deer), and Bovidae (antelopes and relatives).

• **Animals** Animals are multicellular (many-celled) organisms that feed off other organisms. They differ from other multicellular life-forms in their ability to move around (generally using muscles) and their ability to respond rapidly to stimuli.

• **Chordates** Chordates have a dorsal nerve cord—a bundle of nerves running down the back—and a stiff rod called a notochord running along their dorsal (top) side during at least part of their life cycle.

• **Vertebrates** The notochord of a vertebrate transforms into a backbone, or vertebral column, during the development of the embryo. The backbone is made up of a chain of smaller bones called vertebrae, which are made of either cartilage or bone.

• **Mammals** Mammals are warm-blooded vertebrates with mammary glands, which secrete nutritious milk to feed their growing young. All mammals have hairs covering their body, and their lower jaw consists of a single bone. The hinge between the lower jaw and the skull is farther forward than the equivalent jaw hinge of their reptilian ancestors, allowing mammals to chew sideways.

• **Placental mammals** Placental mammals (eutherians) nourish their developing young for an extended period inside the uterus, by means of a temporary organ called a placenta. The placenta joins the developing young to the mother and is formed jointly by the tissue of the embryo and the tissue of the mother's uterus.

▼ *This tree shows the major animal groups to which giraffids belong. Note that some biologists include the cetaceans—whales and dolphins—with the artiodactyls.*

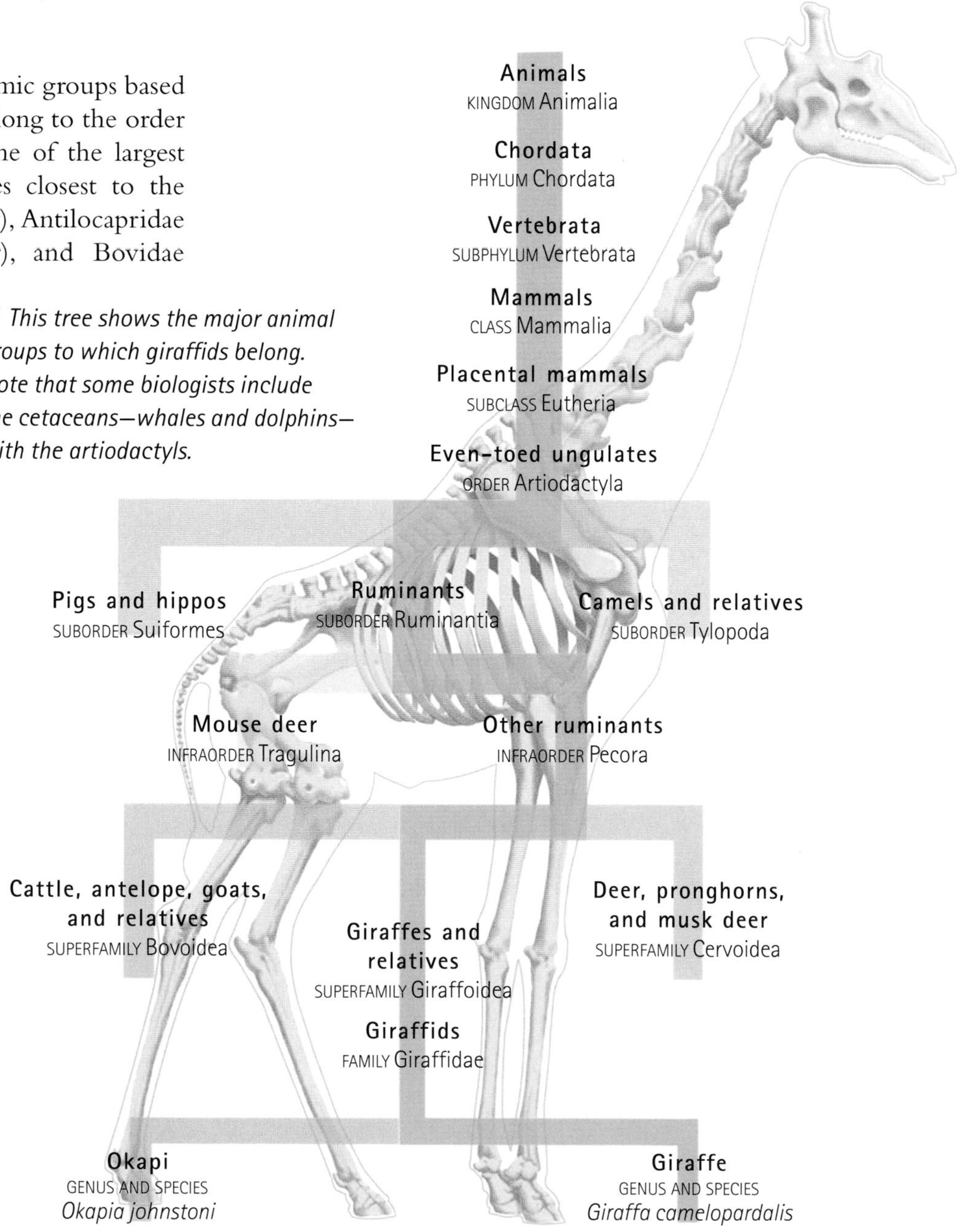

• **Even-toed ungulates** These mammals form the order Artiodactyla. They have an even number of well-developed toes on each foot. The second and fifth toes are usually thinner than the third and fourth toes, and are sometimes tiny or even absent. The animal's body weight is supported along the central axis running between the third and fourth toes. Many species have horns or antlers, and some have enlarged tusklike canine teeth. Artiodactyls range from the tiny mouse deer to the towering giraffe and the 5-ton (4.5-metric ton) common hippopotamus.

• **Deer** These long-legged artiodactyls form one of the larger artiodactyl families. The males of most deer species have antlers, which are usually cast off and regrown each year. Deer skulls have bony platforms that support the antlers. Worldwide, there are at least 41 species of deer, with varying degrees of side-toe reduction.

• **Antelopes and relatives** This diverse family is the largest artiodactyl group, comprising 140 species of antelopes, cattle, goats, and sheep. Most species have horns. All are hoofed, long-legged ruminants, with their weight evenly distributed on two toes on each foot. The third and fourth metapodial bones (metacarpals in the forelimbs and metatarsals in the hind limbs) in each foot are fused to form a single, longer bone called the cannon bone. All bovids have a four-chamber stomach, inside which plant material is digested.

▲ *The skin pattern of giraffes varies between subspecies, but the markings of all giraffes darken with age.*

• **Giraffids** There are just two living species of giraffids—the giraffe and the okapi. Both are tall animals that browse on vegetation; today giraffes occur only in Africa south of the Sahara, although they formerly lived in North Africa, too. Giraffes have a long, narrow head; thin lips; and a long, flexible tongue for browsing. Okapis are smaller and shorter-necked than giraffes, but both species have long, narrow legs and feet without lateral toes; also, their third and fourth metapodial bones are fused to form cannon bones. Giraffes exist in a number of local forms, or subspecies, which have different skin patterns and geographic ranges.

FEATURED SYSTEMS

EXTERNAL ANATOMY Giraffes are large, hoofed mammals with long legs and a very long neck. Their skin is distinctively patterned, and there are between two and five small horns above the eyes. *See pages 358–361.*

SKELETAL SYSTEM The giraffe's skeleton supports its great size and the animal's high-rise browsing lifestyle. *See pages 362–363.*

MUSCULAR SYSTEM Muscles power the giraffe's movements, producing bursts of speed when the animal needs to escape danger. Ligaments extend from the base of the neck to support the massive neck. *See pages 364–365.*

NERVOUS SYSTEM The giraffe has the longest single nerve in the animal kingdom. It runs from the brain to the heart and back again—a distance of around 15 feet (4.5 m). *See pages 366–367.*

CIRCULATORY AND RESPIRATORY SYSTEMS Pressure-reducing vessels offset sudden buildups in blood pressure as the animal bends its long neck. *See pages 368–369.*

DIGESTIVE AND EXCRETORY SYSTEMS Giraffes are ruminants, with a digestive system that enables them to extract the maximum nutrition from their tough food. *See pages 370–371.*

REPRODUCTIVE SYSTEM Male giraffes can determine whether or not a female is ready to mate by tasting chemicals in her urine. *See pages 372–373.*

External anatomy

> **CONNECTIONS**
>
> COMPARE the horns of a giraffe with those of a ***RED DEER***.
>
> COMPARE the coat patterns of a giraffe with those of other large grassland ungulates, such as the ***WILDEBEEST*** or ***ZEBRA***.

With its extraordinarily long neck and high shoulders that slope steeply to its hindquarters, a giraffe resembles a crane on a construction site. In addition to its great height, it is also one of the heaviest land animals: large males can weigh up to 4,200 pounds (1,900 kg). Females are smaller, usually less than half that weight. Compared with other hoofed mammals the giraffe has a relatively short body, but its legs are disproportionately long. The front legs are marginally longer than the hind legs, a feature that contributes to the animal's steeply sloping back. Mature giraffes have hooves as large as dinner plates.

The giraffe and okapi have short **horns** that are fused to the skull. The horns are unique among mammals, consisting of bony cores (called ossicones) covered by skin and fur.
The large **eyes** are protected by thick eyelashes and are set wide on the head, giving the giraffe maximum field of vision.
The **body** appears disproportionately small compared to the neck and legs, with high shoulders sloping steeply down to the tail. The deep chest contains the very large lungs and heart. Enlarged groups of muscles bunched above the shoulders make the front of the animal appear even bulkier.
The giraffe's **tongue** is 18 to 20 inches (46 to 50 cm) long and blue-black. It is extendible and flexible enough to curl around the most nutritious foliage high in trees when the giraffe is browsing. The tongue is also used for grooming.
The giraffe's **markings** vary from one geographical region to another and provide a means of identifying the eight different subspecies. The markings break up the outline of the giraffe and may provide camouflage.
Despite appearances, the hind **legs** are almost as long as the front.
The **foot** is very large—6 inches (15 cm) high in males. Giraffe hooves lack the scent glands that occur in okapi hooves.

▶ Reticulated giraffe
With its long legs, towering neck, and patterned hide, the giraffe is unique and unmistakable.

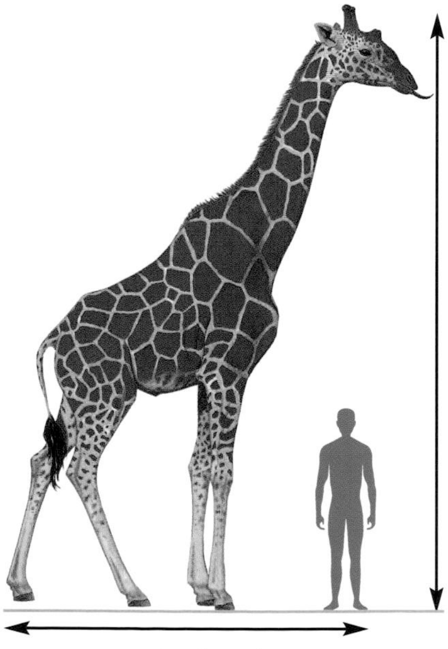

The giraffe's long neck helps it eat leaves that are beyond the reach of other animals. A giraffe can extend its tongue 18 inches (45 cm), curling it around leaves and pulling them toward the mouth. The canine teeth have deep grooves that enable the animal to strip the foliage. The tongue and lips are covered by hard growths called papillae, which are a vital adaptation for feeding on thorny trees. Giraffes have a good sense of smell, their eyes are large, and their vision is excellent. With their unique high-rise vantage point, giraffes have a panoramic view of their surroundings and the best range of vision of any land animal.

Perfect patterns

Giraffes have short, thick fur with intricate patterns of colored patches, which vary from sandy yellow and pale tan to chestnut and almost black, depending on the area in which the giraffe lives and the dominant food types there. Giraffes' striking coloration breaks up their outline, helping conceal them among the trees and bushes of the savanna landscape.

Although no two giraffes' skin patterns are identical, some regional trends are apparent. These provide the basis for the division of the giraffe into a number of subspecies. Eight subspecies are currently recognized, although there may be 12 or more. Among the most striking subspecies are the reticulated giraffe, which has a bold pattern of squares separated by thin white lines that looks like broken paving stones. The Masai giraffe, which sports jagged patterns resembling leaves, is equally striking. Each animal also has its own unique pattern, which enables scientists to study the lives of individual giraffes closely.

A horny head

Both sexes have two to five distinct, bony horns on their head called ossicones. They are covered in skin and, in females, are slender and

▼ *A giraffe's tongue and lips are covered with hard growths called papillae. They enable the giraffe to eat leaves from thorny trees, such as acacias, without being cut by the thorns.*

COMPARATIVE ANATOMY

Skin patterns

Giraffes are patterned with brown patches against a light background. These unusual markings may act as camouflage. In reticulated giraffes, the skin has dark, evenly spaced, boxlike patterns. White spaces between the patches form narrow lines that further break up the animal's shape. This complex pattern may provide excellent camouflage in dry, sunny bush country. In Nubian giraffes the dark patches are darkish-red to chestnut-colored; Masai giraffes have irregular dark patches on a buff-colored background. The exact pattern on a particular giraffe's skin is unique; this may help individuals identify each other. The body of the okapi is much darker than that of the giraffe. The okapi has zebralike stripes on its legs and hindquarters.

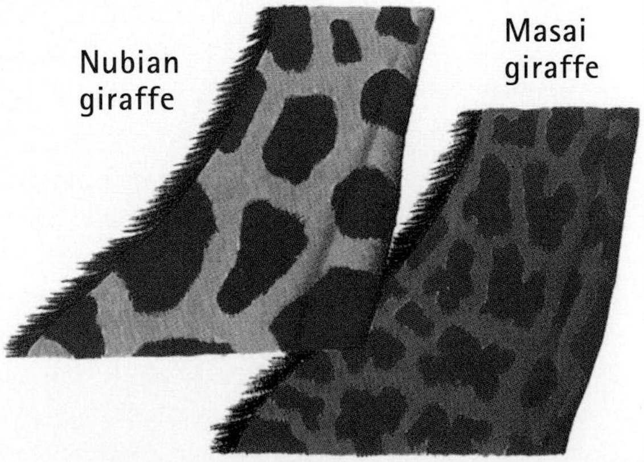

IN FOCUS

Beasts of myth and mystery

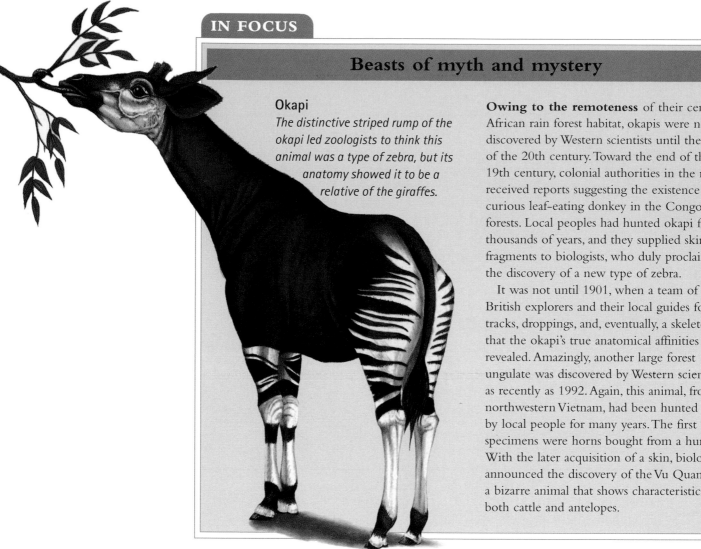

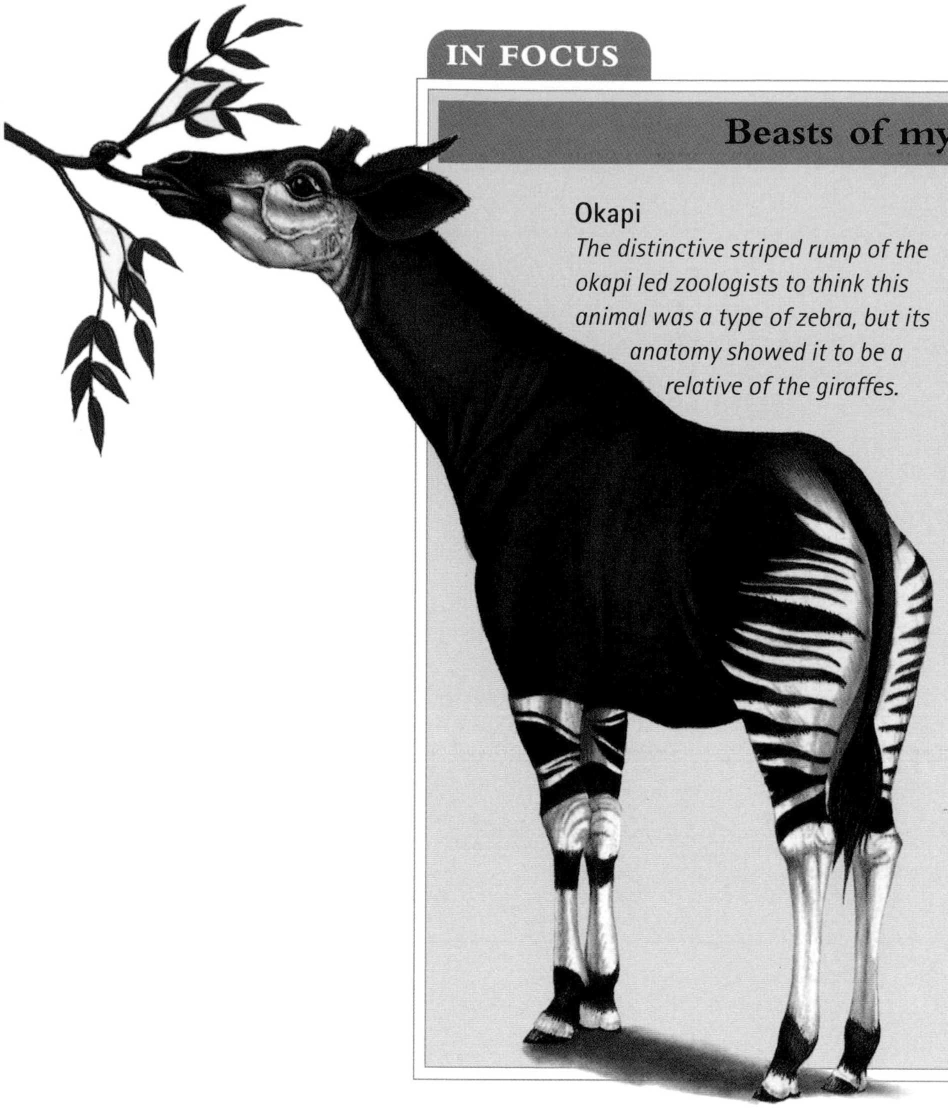

Okapi
The distinctive striped rump of the okapi led zoologists to think this animal was a type of zebra, but its anatomy showed it to be a relative of the giraffes.

Owing to the remoteness of their central African rain forest habitat, okapis were not discovered by Western scientists until the start of the 20th century. Toward the end of the 19th century, colonial authorities in the region received reports suggesting the existence of a curious leaf-eating donkey in the Congo forests. Local peoples had hunted okapi for thousands of years, and they supplied skin fragments to biologists, who duly proclaimed the discovery of a new type of zebra.

It was not until 1901, when a team of British explorers and their local guides found tracks, droppings, and, eventually, a skeleton, that the okapi's true anatomical affinities were revealed. Amazingly, another large forest ungulate was discovered by Western science as recently as 1992. Again, this animal, from northwestern Vietnam, had been hunted by local people for many years. The first specimens were horns bought from a hunter. With the later acquisition of a skin, biologists announced the discovery of the Vu Quang ox, a bizarre animal that shows characteristics of both cattle and antelopes.

tufted with black hair. Males have larger, thicker horns that are bald on top; they grow up to 10 inches (25 cm) long. Males use their horns to spar with one another during fights to establish dominance in the breeding season. The ossicones start off as bumps of cartilage on the forehead, with the cartilage being replaced by bone as the animal ages. This process of bone replacement is called ossification. Eventually the ossicones fuse with the bones of the skull. They remain covered by skin throughout the giraffe's life and continue to grow bigger and thicker in males. Some giraffes also develop a central knob between the eyes, making them five-horned.

A deposition of bony layers onto the skull of males occurs as the animals get older. The head becomes progressively heavier, more clublike, and more angular throughout the animal's life—a trait found in no other mammals.

Long legs and big feet

A giraffe's legs are very long, but despite appearances to the contrary, there is little difference in length between the forelegs and hind legs. The limbs appear inflexible, giving the animal a stiff-legged appearance when walking. The hooves are huge, measuring up to 6 inches (15 cm) across in large males and up to 4 inches (10 cm) across in females. In okapi, the hooves contain scent-secreting glands. These are absent in giraffes.

Legs and hooves are used as formidable weapons when a would-be predator, such as a lion, threatens. The hind legs can give a powerful and direct kick to the rear. The front legs may be employed to deliver either a "chop kick" with the hooves or a bludgeon with the whole straight leg. Either defensive maneuver is effective. One well-placed kick can easily cripple or kill a predator.

EVOLUTION

A shrinking family

Scientists think that giraffids evolved from small, deerlike ancestors around 20 million years ago. Early giraffids were among the first artiodactyls to evolve into large animals and move from dense forests into more open habitats. The expansion of Africa's plains during the Pliocene epoch (2 million to 5 million years ago) triggered a rise in the number of giraffid species, such as *Samotherium boissieri*, a type of grazing okapi. By the start of the Pleistocene, around 2 million years ago, at least seven species of giraffids browsed on the plains of Africa and Asia. They included *Giraffa jumae*, an animal even taller and heavier than the modern giraffe. However, over time the family dwindled. Just two species of giraffids survive, but some relatives disappeared quite recently. *Sivatherium* was a genus of stocky giraffes with two large ossicones on the head and a smaller pair on the muzzle. The last of the genus, *Sivatherium giganteum*, may have become extinct as recently as 8,000 years ago; a Sumerian bronze statuette, which looks very like a *Sivatherium*, suggests that this species may have survived even later in parts of Asia.

▼ *Modern giraffes have a relatively longer neck than their extinct* Sivatherium *cousins, which looked more like okapis.*

Skeletal system

The giraffe's skeleton has to support the animal's body as well as protect delicate internal organs. It also has to allow for movement, often at great speed.

The skull

The upper part of a giraffe's skull is filled with large air canals, or sinuses. The sinuses make the giraffe's skull, especially the female's, surprisingly light for its size. The sinuses are formed by a division of the bones that make up the roof of the skull. As these split in the middle and grow apart, the space between them is covered by a fine layer of bone. The sinuses provide a light but strong platform for the giraffe's ossicones, or horns, and give extra protection to the brain of the male when engaged in fighting rivals.

Limb bones

The forelegs are only slightly longer than the hind legs, but the high dorsal spines on vertebrae at the shoulder make the forelegs appear longer and give the illusion of a steeply sloping backbone. The forelimb bones are generally separate, although in adults some parts of the forelimb skeleton become fused. The radius and ulna (the bones of the lower foreleg) are longer than the humerus (the upper bone). The radius and ulna articulate with the carpal bones, which are equivalent to the human wrist. Two of the metacarpals are unusually long, almost matching the radius in length, and are fused to create the cannon bone typical of many ruminants. There is a pair of

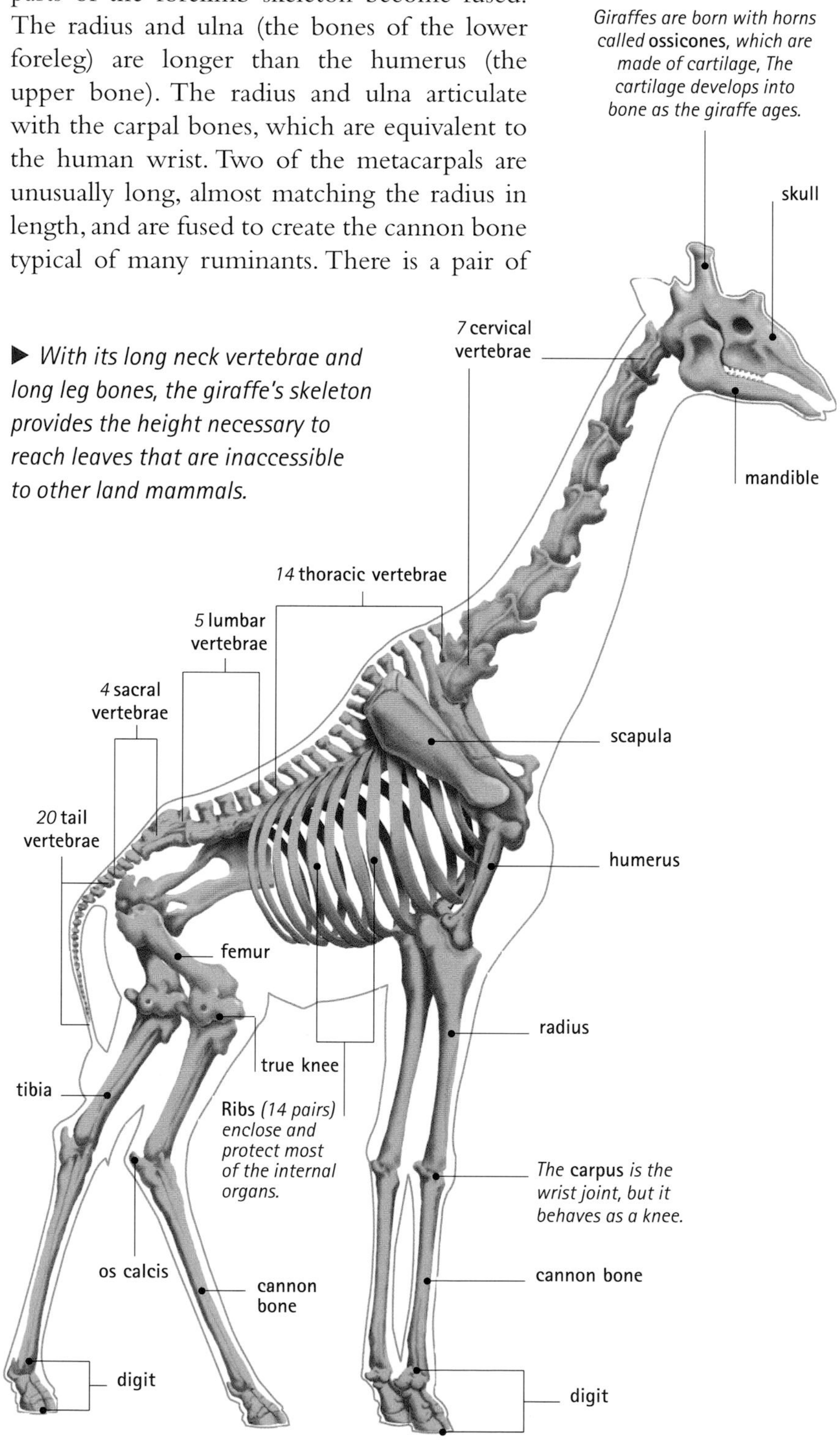

▶ *With its long neck vertebrae and long leg bones, the giraffe's skeleton provides the height necessary to reach leaves that are inaccessible to other land mammals.*

CLOSE-UP

Teeth and gaps

Instead of incisor or canine teeth in the upper jaw, a giraffe has a horny pad against which the lower teeth bite to grind down food. The lower incisor teeth are unusually large, ridged, and broad-crowned. They allow a giraffe to "comb" leaves from the treetops when it is browsing.

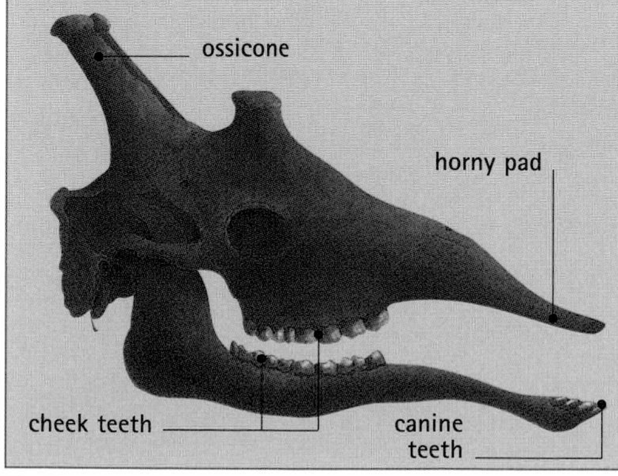

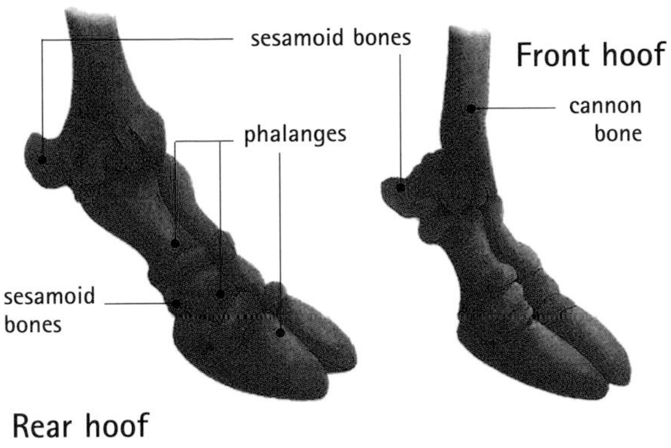

▲ **FOOT BONES**

A giraffe's hoof contains six sesamoid bones (not all visible here), which form attachments for ligaments that help stabilize the legs.

◀ *A giraffe's lower jawbone is called the mandible. It articulates with the animal's skull. The horns, or ossicones, are covered with skin and hair. Male giraffes use these horns in ritualized fights for dominance.*

phalanges, comprising three bones, the first very long and the third slotting into the hoof. Like all ruminants, giraffes have six small bones, the sesamoids, which link a system of ligaments that stabilize the legs. The sesamoids prevent crippling overextension of the legs during running. The humerus articulates with the scapula, or shoulder blade. The giraffe scapula is the longest of any mammal, providing a huge area for muscle attachment.

The femur (thighbone) of the hind limb articulates with the pelvic girdle, and the tibia (shinbone) articulates with the metatarsals (equivalents of the foot bones) via the tarsal (ankle) bones. Some of these tarsal bones are fused, leaving four bones in giraffes and three bones in okapis.

The spine

The giraffe's backbone, or spine, contains around 50 vertebrae, the same number as cattle. Remarkably, there are just 7 neck vertebrae, the same as in almost all other mammals. The spine also comprises 14 thoracic, 5 lumbar, 4 sacral, and 20 tail vertebrae. The thoracic bones have long spines on the dorsal (upper) surface to which the muscles and ligaments of the enormously elongated neck are attached. All the thoracic vertebrae bear a pair of ribs, of which 7 are fused to the curved sternum (the breastbone) and 6 to a support made of cartilage. The ribs form a protective basket around most of the internal organs. The first thoracic vertebra bears particularly long dorsal spines. They form an attachment point for many of the powerful neck muscles.

The neck vertebrae are very large—more than 11 inches (28 cm) long. Unusually, a giraffe's neck bones are connected by a ball-and-socket articulation; among many other examples, this arrangement also occurs in the neck bones of camels and in the human pelvis. This gives the neck great flexibility. The neck vertebrae are rounded at the anterior (front) end and slot neatly into the concave posterior (rear) end of the next bone.

COMPARATIVE ANATOMY

Articulated for flexibility

The type of ball-and-socket arrangement that occurs between a giraffe's neck vertebrae is called an opisthocoelous joint. It is similar to the articulation found in snake vertebrae. In these reptiles, a fluid-filled capsule encloses both ends of each vertebra and connects adjacent vertebrae, allowing them to slide freely across each other. This gives the snake's backbone its amazing flexibility. In giraffes, a different system provides the flexibility: the ends of the vertebrae are linked by layers of ligaments rather than fluid-filled capsules.

Muscular system

CONNECTIONS

COMPARE the neck muscles of a giraffe with those of a short-necked animal such as a ***HIPPOPOTAMUS***.

COMPARE the long, thin leg muscles of a giraffe with the strong, broad muscles on the forelegs of a digging animal such as a ***GIANT ANTEATER***.

A giraffe's muscular system provides the pulling power needed to move its huge body around. The muscular system consists of three different types of muscle tissues: skeletal, cardiac, and smooth. Each can contract to allow body movement and functions. Some muscles are voluntary; the giraffe controls voluntary muscles when it requires a specific action or movement, such as moving its legs or neck. Involuntary muscles are those that contract automatically, such as the heart muscles and intestinal muscles.

Muscle power

Giraffes, along with other ruminants, have no clavicle (collarbone). Instead, the shoulder blades are deeply embedded in thick muscle. Several enormous muscles support the front end of a giraffe's torso: the cephalo-humerals, deltoids, triceps, and latisimus dorsi muscles. Between these muscles and the leg bones is an area of tough, elastic cartilage that helps the animal run efficiently.

IN FOCUS

Tests of neck strength

A giraffe's neck is held upright by a strong, elastic tissue called the ligamentum nuchae. This extends from as far back as the lumbar vertebrae and connects regularly all along the spine before running up the neck in two tightly bound halves to join the back of the skull. Young bull giraffes take part in "necking" contests, ritualized fights in which both males slowly entwine their necks, push backward and forward against each other, and butt heads. These wrestling matches may last for 30 minutes or more and provide the young males with an opportunity to develop and test their neck muscles. These contests, however, become more serious in adults, when they determine which males get to mate with the local females.

IN FOCUS

Squeezing power

The tongue is useful for grabbing food. The tongue is also vital in the process of rumination: the rechewing of food to help maximize digestion. A bolus of food is regurgitated from the rumen (front part of the stomach) into the mouth, where it is chewed. The tongue's muscular strength is used to apply pressure on the bolus, squeezing the moisture out of the food parcel. Then the giraffe swallows it for a second time, and it is further digested.

The leg muscles are concentrated in the upper leg; the lower leg mainly contains long tendons that facilitate movement of the hooves. This structure acts as another energy-saving device: at high speed, only a little effort is needed to move the muscles near the leg's pivot. This small movement is translated into a wide arc of movement of the hoof.

A giraffe has two modes of movement, or gaits—an ambling walk and a gallop. As it walks, the legs on one side of the body move together, followed by their partners on the opposite side. Body weight is therefore supported alternately on the left and right sides. In the galloping gait, forelegs and hind legs work in pairs: the front pair first, then the back. The hind hooves swing up like a pendulum and are placed in front of the fore hooves. This gait enables the giraffe to reach speeds of up to 38 miles per hour (60 km/h). A giraffe maintains its balance because the neck moves back and forth simultaneously with the legs.

Lips and tongue

A giraffe's long, black, muscular tongue reaches up to 18 inches (46 cm) in length. It is also prehensile, acting like a grasping hand as it wraps around branches and leaves. The tongue

▲ **GALLOPING**

When a giraffe is galloping, its forelegs and hind legs work in pairs. The hind legs are brought in front of and outside the forelegs. The forelegs are then raised off the ground, and the hind legs provide the push for the giraffe to move forward.

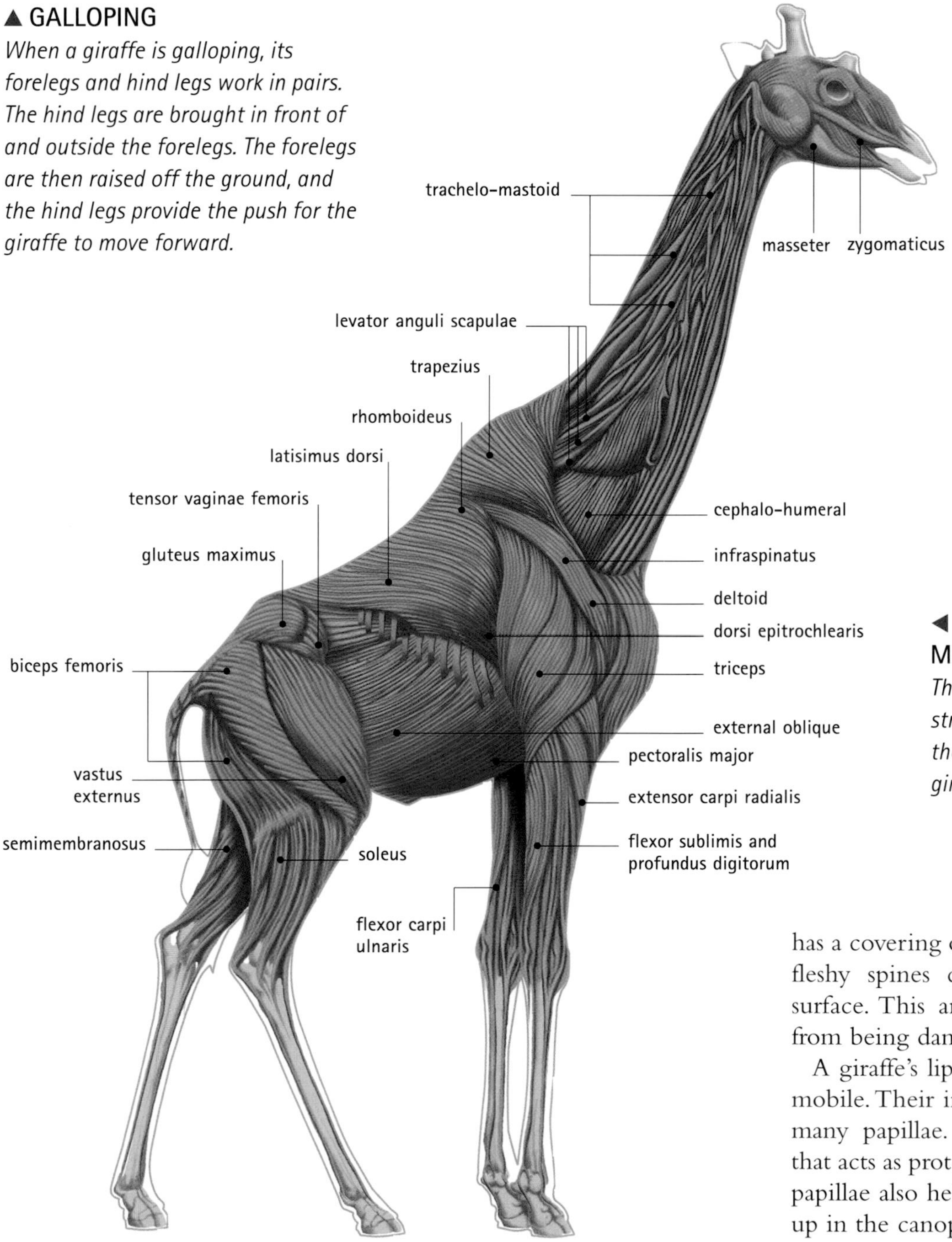

◀ **SUPERFICIAL LAYER MUSCLES**

The giraffe's muscle system must be strong enough to support and move the long neck as well as enable the giraffe to run swiftly from predators.

has a covering of tough skin with many small, fleshy spines called papillae on the upper surface. This arrangement keeps the tongue from being damaged by sharp thorns.

A giraffe's lips are also highly muscular and mobile. Their inner surfaces are covered with many papillae. They create a rough surface that acts as protection against spiny foliage; the papillae also help the giraffe reach leaves high up in the canopy.

Nervous system

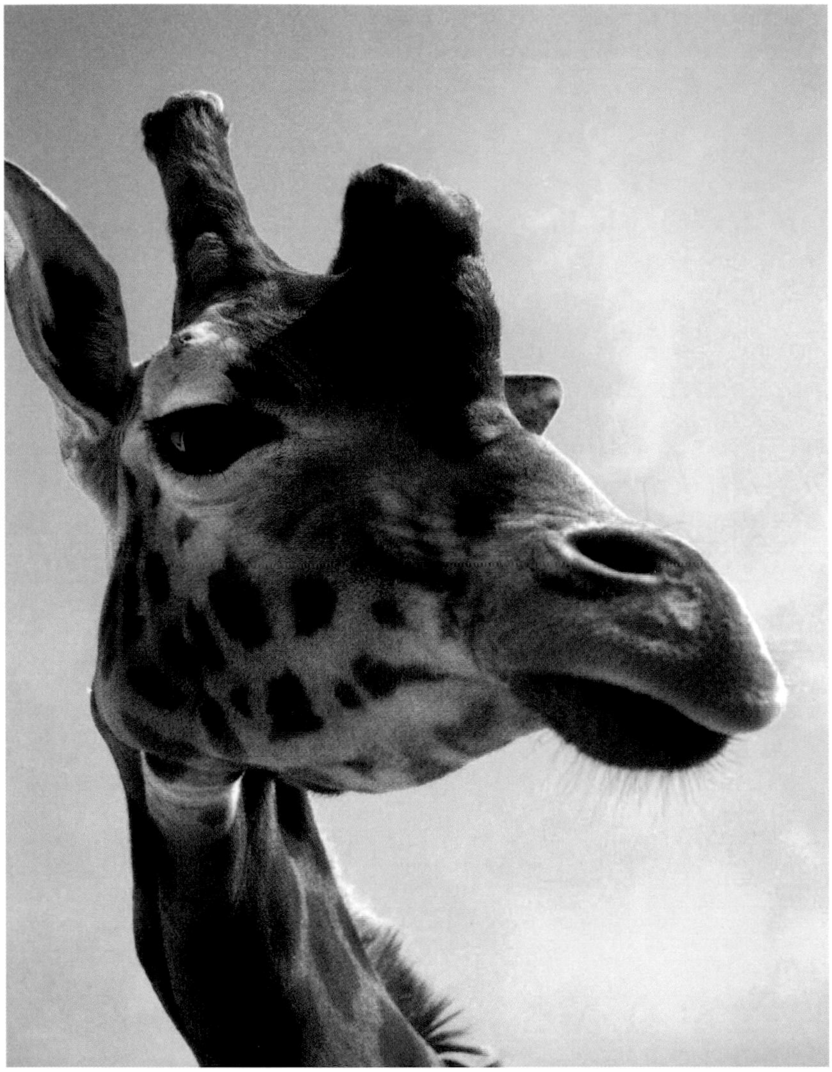

◀ *A giraffe's highly movable lips are covered with sensitive hairs. Giraffes also have a good sense of smell, good hearing, and keen eyesight.*

The nervous system is a vast network of cells carrying information through the body by means of chemical and electrical signals. The nervous system is divided into two main sections: the central nervous system (CNS) and the peripheral nervous system (PNS). The CNS comprises the brain and the spinal cord; the PNS includes nerves that transmit signals from the sense organs to the CNS and vice versa.

The nervous system is made up of specialized cells called neurons. They bear long processes, or dendrites. Many neurons have one or more extra-long processes called axons that allow long-distance communication. These neurons are the longest cells in an animal's body, and giraffes have some of the longest nerve cells in the animal kingdom.

There are three main types of neurons: sensory neurons, motor neurons, and interneurons. Sensory neurons connect sense organs such as the eyes and ears to the CNS. Motor neurons carry signals from the CNS to muscles. Interneurons, which occur only in the CNS, connect sensory and motor neurons together.

Voluntary and involuntary

The PNS is divided into two distinct systems: the somatic nervous system and the autonomic nervous system. The somatic system controls voluntary actions, such as walking. The autonomic system controls involuntary body processes over which the animal has no conscious control, such as the heartbeat. The autonomic nervous system also triggers output from certain glands around the body.

Neck and throat

A giraffe's neck has eight pairs of nerves along its length. Most notable of these is the giraffe's laryngeal nerve—the longest nerve in the animal kingdom—which measures around 15

IN FOCUS

Communicating over distance

Giraffes and okapis were once thought to be virtually silent. However, they can communicate vocally over enormous distances. They make long-distance calls using infrasound. A giraffe typically lowers its chin, then quickly raises it to produce the sounds. Infrasonic signals are sound waves produced at such a low frequency that they cannot be heard by the human ear. Infrasound provides a good way to warn other giraffes of danger while remaining hidden. Infrasonic signals are very difficult to pinpoint, so any predators that could hear infrasound would find it hard to locate their source.

feet (4.5 m) long. It begins at the brain and runs down the length of the neck. It crosses over a blood vessel at the top of the heart before looping back up the neck to the larynx.

Vision

Giraffes depend on vision. They have excellent eyesight, enabling them to locate both food and distant predators from their lofty position above the savanna. A giraffe's eyes are proportionally larger than those of other ruminants, such as deer and cattle. The positioning of the eyes on the sides of the head also gives the giraffe superior peripheral (sideways) vision. A giraffe has good color vision. This helps giraffes recognize each other and remain in visual contact with other giraffes over long distances.

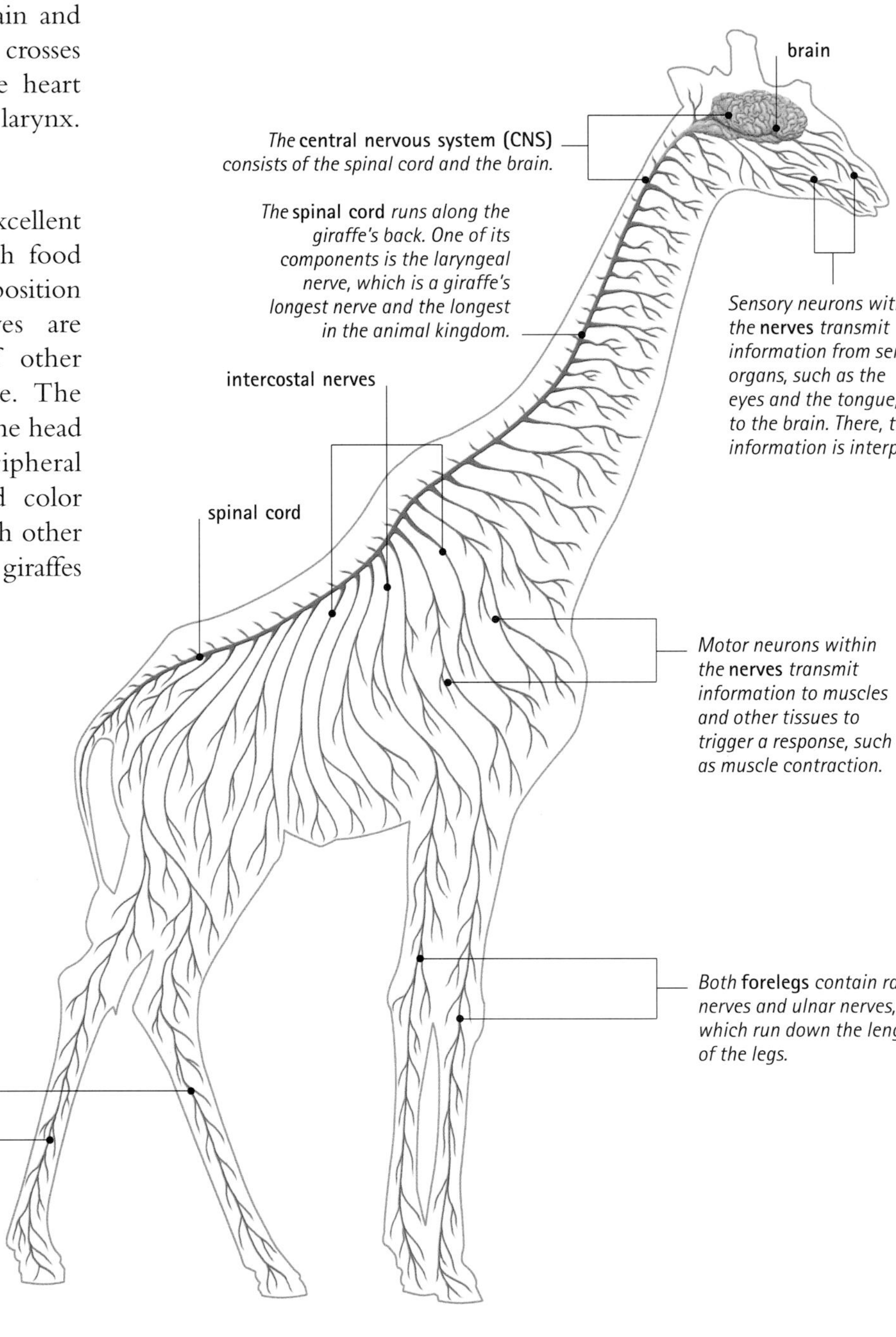
brain
The **central nervous system (CNS)** *consists of the spinal cord and the brain.*
The **spinal cord** *runs along the giraffe's back. One of its components is the laryngeal nerve, which is a giraffe's longest nerve and the longest in the animal kingdom.*
intercostal nerves
spinal cord
Sensory neurons within the **nerves** *transmit information from sense organs, such as the eyes and the tongue, to the brain. There, the information is interpreted.*
Motor neurons within the **nerves** *transmit information to muscles and other tissues to trigger a response, such as muscle contraction.*
Both **forelegs** *contain radial nerves and ulnar nerves, which run down the length of the legs.*
Both **hind legs** *have peroneal nerves and tibial nerves, which run along their entire length.*

▶ *A giraffe's nervous system is similar to that of other mammals, with nerves of the peripheral nervous system (PNS) branching in pairs from the spinal cord.*

COMPARATIVE ANATOMY

The brain

Giraffes have a small brain, weighing just 1.5 pounds (680 g). This represents 0.05 percent of the animal's body weight, a little less than that of a cow. By contrast, a dolphin's brain makes up 0.8 percent of its total weight, and a human's brain accounts for 2 percent. The giraffe's brain is actually smaller than expected for an animal of its size. This disproportion may be related to the length of a giraffe's neck. Much more energy would be used to supply enough oxygen to a larger brain at the end of such a long neck.

Circulatory and respiratory systems

Like all vertebrates, giraffes have a closed circulatory system—blood is pumped through a system of arteries, veins, and capillaries. As oxygen and nutrients diffuse from the blood into the tissues, waste materials move into the blood to be taken away.

Pressure regulation

Giraffes have a four-chamber heart with two atria and two ventricles. Oxygenated blood from the lungs and deoxygenated blood returning from the body are kept separate; the four chambers ensure efficient transport of oxygenated blood to the body's organs. The giraffe's large heart weighs more than 24 pounds (10 kg) and is among the strongest in the animal kingdom. That is because almost double the normal amount of pressure is needed to pump blood 10 feet (3 m) up the neck to the brain.

The heart beats around 150 times per minute. This rate is unusually high for an animal of such a size; usually, the larger the animal, the slower its heartbeat. Scientists have calculated that blood leaves the giraffe's heart at a pressure of up to 6 pounds per square inch (40 kilopascals), which is the highest blood pressure of any living animal. The red blood cells are small, but there are twice as many of them per unit of volume as in human blood. This combination of small size and very high density allows the cells to absorb oxygen quickly and efficiently.

▶ ARTERIAL SYSTEM

The most important arteries are shown on the diagram. Arteries carry oxygenated blood to all parts of the body, while veins (not shown) carry deoxygenated blood back to the heart.

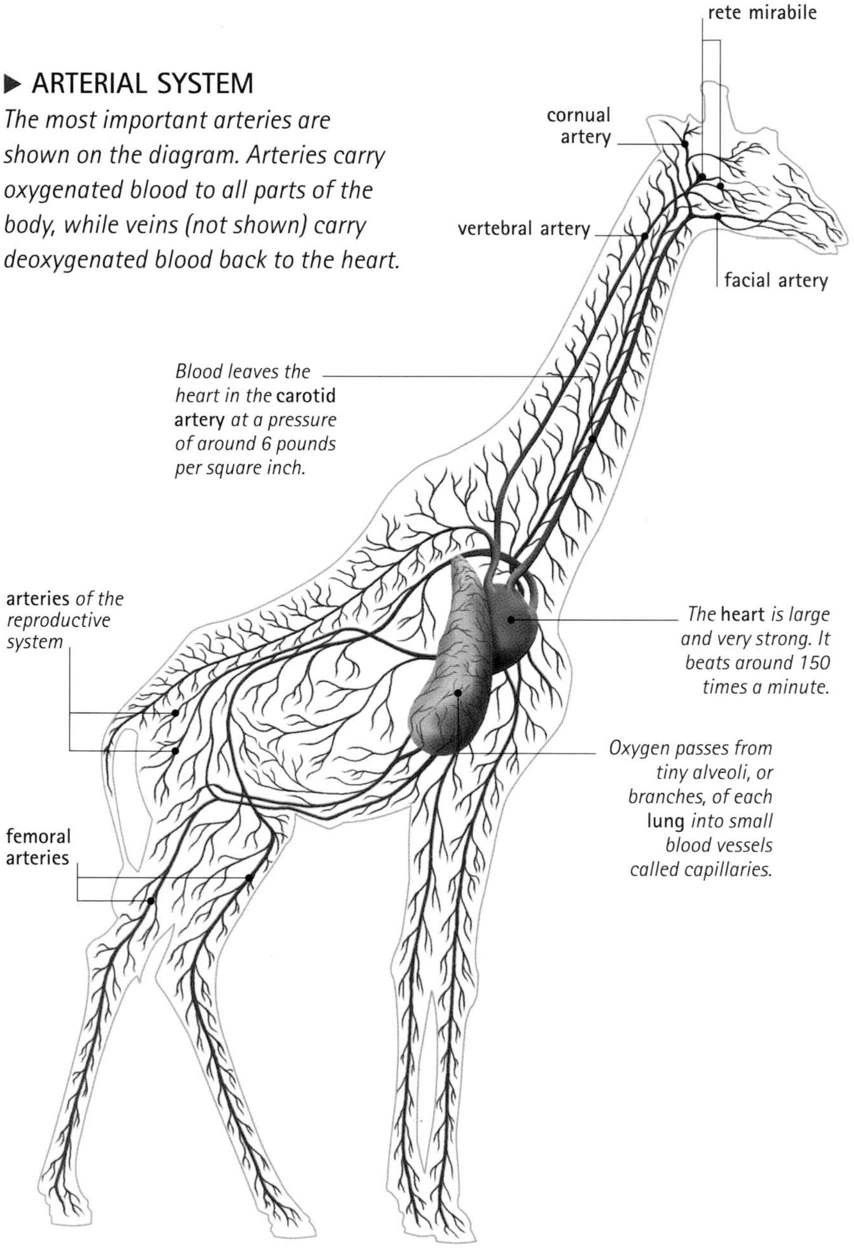
rete mirabile
cornual artery
vertebral artery
facial artery
Blood leaves the heart in the carotid artery at a pressure of around 6 pounds per square inch.
arteries of the reproductive system
The heart is large and very strong. It beats around 150 times a minute.
Oxygen passes from tiny alveoli, or branches, of each lung into small blood vessels called capillaries.
femoral arteries

IN FOCUS

Arterial protection

To cope with massive blood pressure surges, a giraffe's arterial walls are much thicker than those of other animals. The vessels are also deeply embedded for protection. The main arteries and veins in the legs lie in deep bony furrows beneath the tendons and do not come close to the surface. Fluid surrounding the cells of the body is kept at a high pressure; that is largely achieved by the extremely thick skin, which is stretched tightly over the body. Scientists studied giraffe's skin when they were developing suits for astronauts and the pilots of fighter planes. These suits prevent blood from rushing to the legs and causing the pilots to pass out during rapid ascents.

Another unusual feature of a giraffe's circulatory system is its ability to maintain a far lower pressure at the brain of around 1.7 pounds per square inch (12 kilopascals), which is no higher than in other large mammals. This regulation is essential when a giraffe lowers its head to eat or drink. Without it, blood would rush down the long neck into the brain, causing the blood vessels to burst.

The rete mirabile

Blood pressure in the brain is controlled by a web of tiny blood vessels located at the base of the brain. This is called rete mirabile and is formed by many subdivisions of the carotid arteries. This web is crucial for maintaining blood flow to a giraffe's brain at the right pressure. The walls of the rete mirabile blood vessels are elastic and can expand to cope with the increase in blood pressure when the giraffe lowers its head. The walls can also contract again when the head is raised.

The carotid artery is a single strand along most of the neck but divides into an internal and an external branch near the head; the external branch forms the rete mirabile. The carotid artery is also linked by a small branch to another artery, the vertebral. This runs down the neck, with many branches supplying the muscles. The vertebral artery acts as a further safeguard, draining off much of the blood before it even reaches the rete mirabile when the giraffe's head is lowered. Retia mirabilia occur in the carotid arteries of many other artiodactyl mammals. They may have an important role in dissipating heat.

CLOSE-UP

How giraffes breathe

A giraffe's unusually long neck poses a great challenge to efficient breathing. A giraffe's windpipe is more than 5 feet (1.5 m) long, yet it is only around 2 inches (5 cm) in diameter. The tube usually contains around 0.8 gallon (3 l) of air. A giraffe inhales a lot of air that is never used for respiration, so the windpipe is always filled with a mix of inhaled and exhaled air, and oxygen levels are correspondingly low. To overcome this problem, a giraffe has to breathe much more regularly than would be expected for an animal of its size. A giraffe takes more than 20 breaths a minute when resting, compared with around 12 in humans and 10 in elephants.

▼ PRESSURE CONTROL

When a giraffe drinks, it has to lower its head far below its heart, so one would expect the blood pressure to increase. A network of elastic blood vessels called the rete mirabile expands to lower the pressure of blood entering the brain. Without this elasticity the blood vessels in the giraffe's head would burst.

Digestive and excretory systems

CONNECTIONS

COMPARE a giraffe's four-chamber stomach with that of a carnivore such as a ***LION***.

COMPARE a giraffe's muscular esophagus with that of a ***RAT***. The muscles of a giraffe's esophagus need to be very strong to push food up from the stomach when it is regurgitated.

Giraffes are highly selective when it comes to diet. They browse on trees, especially acacias, and eat their leaves, buds, and young shoots. Acacias are well-protected, with sharp thorns, and ants live inside the branches that protect their homes with staggering ferocity. With their tough mouthparts, reinforced tongue, and thick skin, giraffes can overcome these defenses. As well as essential fatty acids, acacia also contains a high percentage of water, providing giraffes with much of their required daily liquid intake, so they have to seek out water holes only occasionally. The giraffe's esophagus is very long and muscular, and connects to the large stomach. The esophagal muscles are used not only to swallow food but also to push it back up the throat when it is regurgitated for further chewing.

Up and down

A giraffe's stomach has four chambers, as is typical of many ruminants. After being swallowed, a giraffe's briefly chewed food is passed into the first stomach chamber, the rumen. The food is softened there, then regurgitated back into the mouth as a ball of chewed-up food called a bolus. There it is

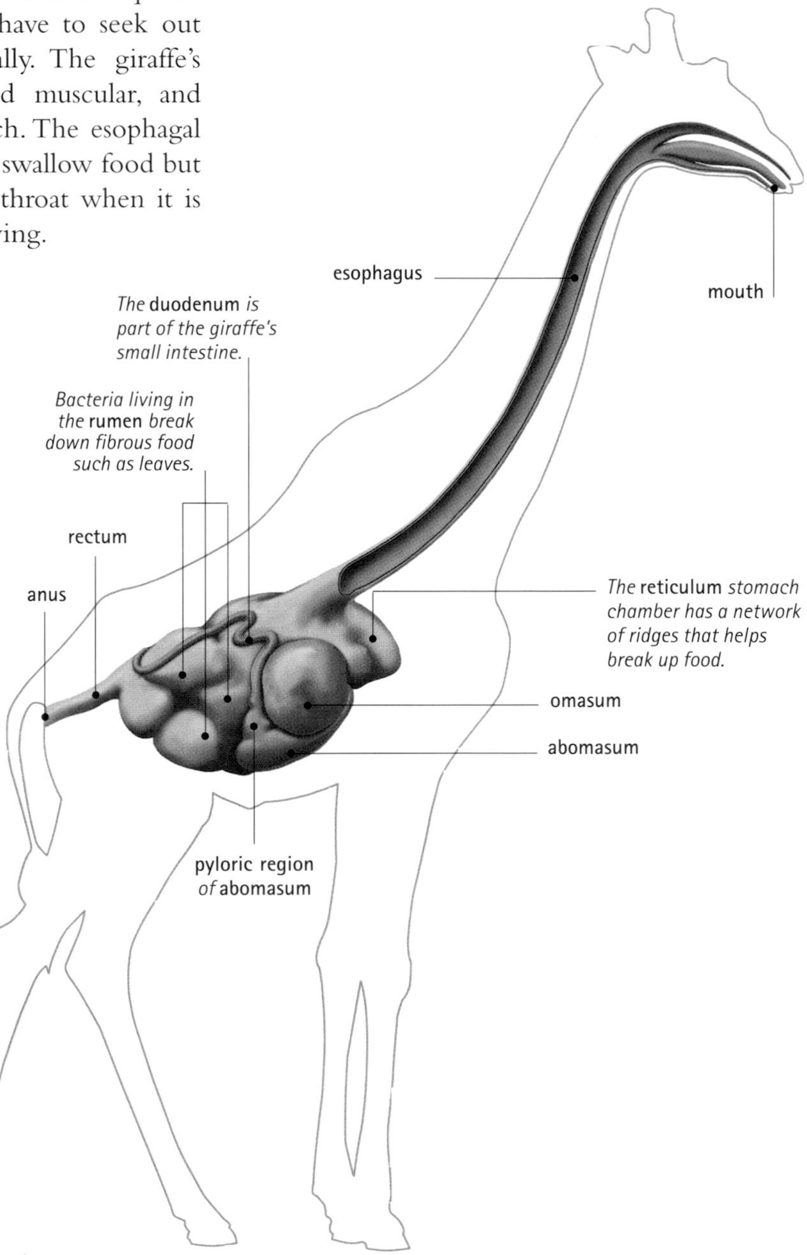
esophagus
mouth
The duodenum *is part of the giraffe's small intestine.*
Bacteria living in the rumen *break down fibrous food such as leaves.*
rectum
anus
The reticulum *stomach chamber has a network of ridges that helps break up food.*
omasum
abomasum
pyloric region *of* abomasum

▶ FOUR STOMACHS
The giraffe's stomach is divided into four parts: reticulum, omasum, abomasum, and rumen. This arrangement enables the giraffe to extract the maximum nutrition from its leaf diet. In the diagram most of the intestines are hidden by the stomachs.

IN FOCUS

Piling on the pounds

Female giraffes usually spend around 55 percent of their day feeding. When food is plentiful they quickly build up huge fat reserves, depositing the fat first around organs such as the heart and kidneys, and also in the tissues that support the intestines and lungs. Reserves are eventually laid down just beneath the skin in a layer of thick white fat. These fat reserves are essential for maintaining good health during pregnancy, helping give the newborn calf a head start in life.

IN FOCUS

A lost organ

Giraffes have kidneys, a pancreas, and a spleen that are similar to those of other ruminants, but the liver is surprisingly small and compact. Giraffes lack a gallbladder, an organ that stores and releases bile; deer and some other ruminants also lack this organ. Bile helps emulsify fats in other organisms, permitting their digestion. The ancestors of these ruminants had a gallbladder, but lost it as they evolved. A tiny gallbladder is present in an unborn giraffe while it is developing in the uterus, but this has disappeared by the time of birth.

subjected to further chewing. The food is swallowed again and forced into the reticulum. This second chamber contains a network of ridges that forms a honeycomb-like pattern. Digestion takes place in the third and fourth stomach chambers, the omasum and the abomasum.

Digestion in the stomach

During digestion, solid food flows slowly through the rumen while water extracted from the plant materials flows through rapidly. This flow of water helps flush the solid food downstream. Bacteria begin to act on the plant matter in the rumen, and fermentation begins. The food is broken down and reduced to ever smaller sizes. The rumen's contractions constantly flush lighter solids back upward, while the smaller, thicker materials are pushed into the reticulum. From there food particles are ejected, floating in a liquid thick with bacteria, into the omasum. Some fatty acids may be absorbed through the walls of this chamber before the food passes into the abomasum—the giraffe's true stomach. This fourth chamber functions in much the same way as a nonruminant mammal's stomach, secreting acids to break down food. Unlike nonruminant stomachs, the abomasum secretes an enzyme called lysozyme, which breaks down bacteria, which is essential, considering the large numbers that collect there.

▲ A giraffe pulls leaves from a tree. The food will pass to the stomach along the esophagus. In a process called peristalsis, muscles in the esophagus force the food down. Peristalsis also forces food back up for rechewing.

The intestines

Digested food is absorbed through the walls of the intestines. A giraffe's intestines are longer than those of other ruminants and may be as long as 280 feet (85 m). The small intestine is tightly coiled and cushioned by a mass of elastic tissue to prevent it from pressing against the surrounding organs. A sheet of muscle called a diaphragm further separates the intestines from the heart and lungs.

Reproductive system

Male, or bull, giraffes produce sperm in organs called testes, which largely consist of twisted spaghetti-like tubes called seminiferous tubules. The female produces eggs in her ovaries. These become mature during estrus, the period when mating takes place. There is little courtship between the sexes. Bulls, however, battle for access to females in estrus. Bulls stand side by side and swing their necks, striking each other with their heads.

Mating is brief. After testing the female's receptivity by tasting her urine, the bull may nudge her gently or attempt to rest his neck across her back. The giraffes circle each other before the bull mounts the female, sliding his forelegs along the female's flanks and propping himself against her.

Fertilization and development

Fertilization of the egg by sperm occurs in the reproductive tract of the female. Female giraffes and okapi have a bicornate uterus. This consists of two "horns," which extend from the cervix (neck of the birth canal) to each of the fallopian tubes.

The fertilized egg develops into an embryo. Giraffes are placental mammals; unborn young receive nourishment from the female through a structure called the placenta, which connects to the young by the umbilical cord. This arrangement allows nutrition and oxygen to enter the developing calf, and waste to move in the opposite direction. It also allows the female to pass on antibodies, which will help the calf fight disease.

▶ A giraffe's reproductive organs are similar to those of other mammals. The male has two testes, which produce sperm. The sperm fertilize the female's eggs, which are produced in the ovaries.

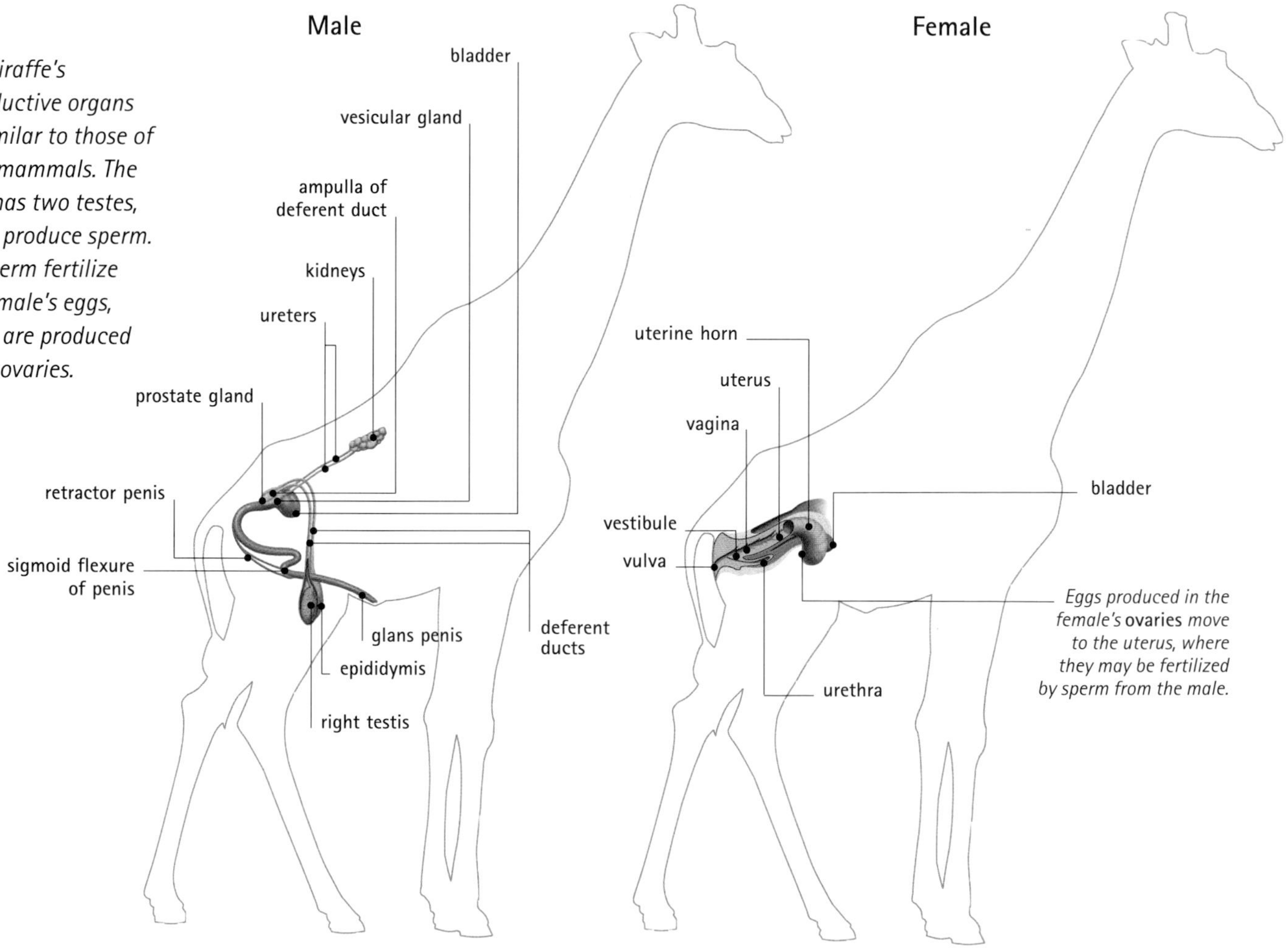

Eggs produced in the female's **ovaries** *move to the uterus, where they may be fertilized by sperm from the male.*

IN FOCUS

Surviving the fall

The long neck of a baby giraffe is fragile. If the calf emerged from the female headfirst, as occurs in humans and other mammals, the neck would risk being broken as the rest of the body fell on top of it. For that reason, giraffes are born feet first. The feet help break the animal's fall. Newborn giraffes need to be tough to survive this fall. This partly explains why they are precocious (well developed at birth). Precocious young are also better able to escape predators and other hazards during the dangerous first few days of life.

▲ *Newborn giraffes are around 6 feet (1.8 m) tall and are often on their feet within 20 minutes of birth. The young giraffe grows swiftly, reaching adulthood in just four years. During this time the neck grows from being one-sixth to one-third of the giraffe's total height. After four years the females are ready to breed, but bulls do not usually breed until they are around seven years old.*

The hazards of birth

Gestation, or pregnancy, is long in giraffes, lasting around 14.5 months—one of the longest in the animal kingdom. The female gives birth to one calf, which she delivers while standing, since she is unable to squat. The newborn calf drops to the ground from a height of 5 feet (1.5 m); it must be robust to survive such a fall. Within two months of giving birth, the female is in estrus again and is able to mate once more. Giraffes usually give birth at intervals of 20 to 23 months.

Mother's milk

A female giraffe has two or four teats. The milk is concentrated and highly nutritious, around five times richer in proteins than cow's milk. However, after a few weeks the protein level in the milk drops by 50 percent, and the fat content is reduced by two-thirds. In contrast, the sugar content (in the form of lactose) doubles at this time. Giraffe calves suckle for between 10 and 16 months, but the young giraffes are able to ruminate solid food after four months.

STEVEN SWABY

IN FOCUS

Dangerous days

The first few months of a giraffe's life are by far the most dangerous. Between 50 and 75 percent of calves fall prey to lions or hyenas during this time, despite rigorous protection from their mothers. Young giraffes rely on their patterned coat to camouflage them; they usually keep a low profile by crouching out of sight in tall grass. Their rich diet helps them quickly develop the muscular power to outrun predators such as lions. However, until they are much older giraffes cannot sustain the speed necessary to outpace more persistent predators such as hyenas over longer distances.

FURTHER READING AND RESEARCH

Nowak, Ronald M. 1999. *Walker's Mammals of the World*. Johns Hopkins University Press: Baltimore, MD.

Vaughan, Terry A. 1999. *Mammalogy*. Brooks/Cole: Belmont, CA.

Gray whale

ORDER: Cetacea SUBORDER: Mysticeti
FAMILY: Eschrichtiidae SPECIES: *Eschrichtius robustus*

The gray whale is a slow-swimming marine mammal that lives in coastal waters of the North Pacific. It migrates between its winter breeding grounds in tropical waters and its summer feeding grounds in polar waters. Gray whales and others of their suborder have a filtering device in the mouth called baleen. They use it to sieve crustaceans and other invertebrates from the seabed.

Anatomy and taxonomy

Biologists categorize all organisms into groups based partly on anatomical features. The gray whale differs enough from other whales to merit placement in a family of its own, the Eschrichtiidae. This family is part of the suborder Mysticeti, the baleen whales, which includes 13 other species.

- **Animals** Animals, are multicellular (many-celled) and gain their food supplies by consuming other organisms. Animals are able to move from one place to another (in most cases, using muscles).

- **Chordates** At some time in its life cycle a chordate has a stiff, dorsal (back) supporting rod called the notochord that runs along most of the length of its body.

- **Vertebrates** The vertebrate notochord develops into a backbone made up of units called vertebrae. The vertebrate muscular system that moves the head, trunk, and limbs consists primarily of muscles that are arranged in a mirror image on either side of the backbone.

- **Mammals** Mammals are warm-blooded vertebrates that have hair made of keratin. Females have mammary glands that produce milk to feed their young. The mammalian lower jaw is a single bone that hinges directly to the skull, and the middle ear contains three tiny bones.

- **Placental mammals** Placental mammals, or eutherians, nourish their unborn young through a placenta, a structure that forms in the mother's uterus during pregnancy.

- **Cetaceans** Members of this group are supremely adapted for life in water, where they spend their entire life. Cetaceans are streamlined like fish. This helps minimize drag as the animal swims. Cetaceans differ from the other

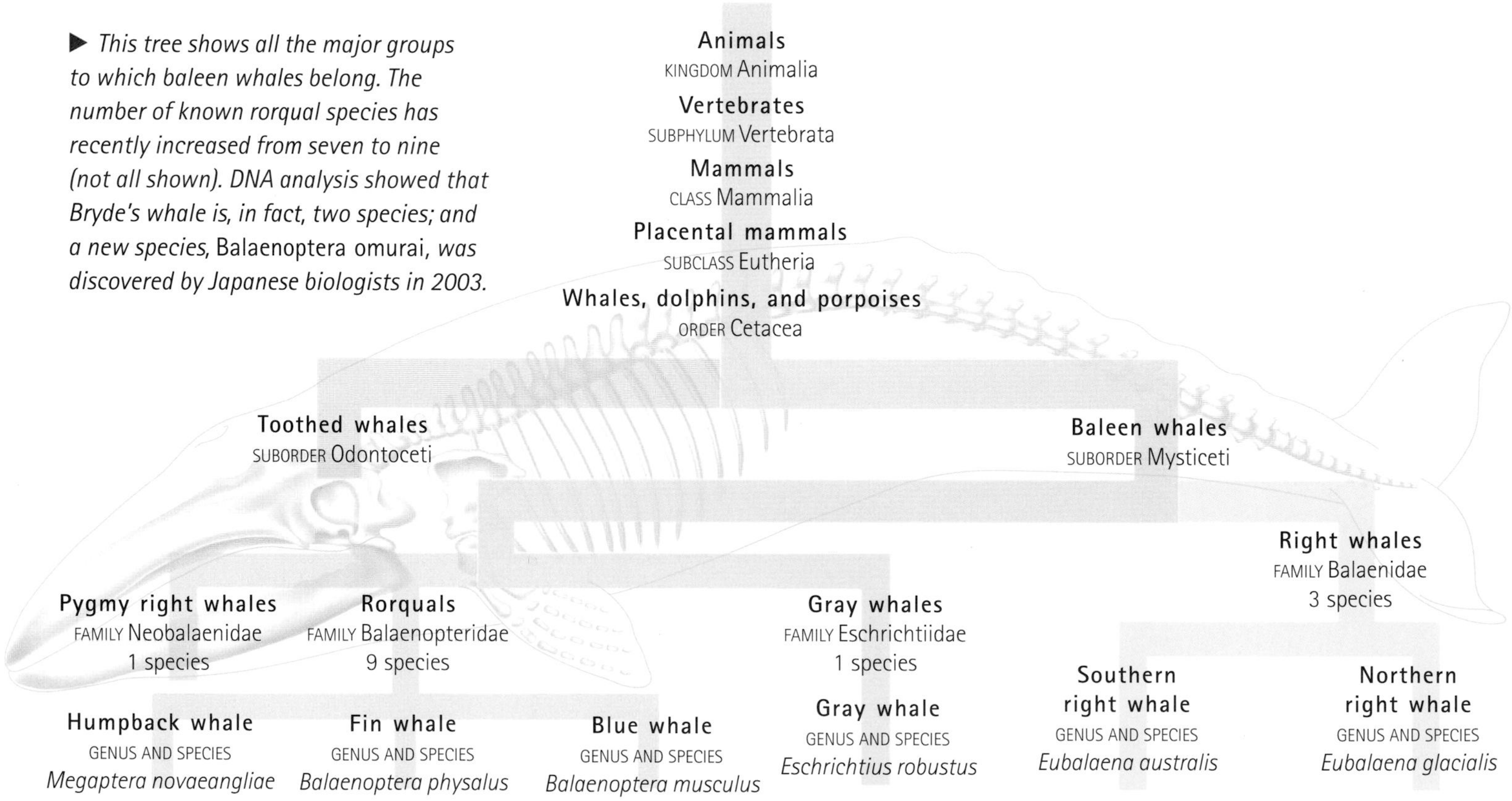

▶ *This tree shows all the major groups to which baleen whales belong. The number of known rorqual species has recently increased from seven to nine (not all shown). DNA analysis showed that Bryde's whale is, in fact, two species; and a new species,* Balaenoptera omurai, *was discovered by Japanese biologists in 2003.*

major marine mammal groups—sea cows, seals, and sea lions—in many ways. For example, cetacean nostrils have moved from the front of the head to the top. This enables easy breathing at the sea surface. The nostrils exit through one or two blowholes. Like sea cows, cetaceans have paddlelike tails, their forelimbs form flippers that lack visible digits, and they have no functional hind limbs.

• **Toothed whales** In most of the 73 or so species of toothed whales, the jaws extend into a beaklike snout armed with teeth. The forehead bulges upward, enclosing the melon, a fat-containing structure that focuses sound waves. This enables the whale to use sound to visualize its surroundings, a process called echolocation. Toothed whales breathe through a single blowhole.

• **Baleen whales** There are 14 species of baleen whales. They include most of the larger whales and the largest whales of all, the fin and blue whales. Instead of teeth, baleen whales have thin, flexible plates of baleen hanging from their upper jaw. Whales strain fish or shrimplike crustaceans from the water with their baleen. All baleen whales have two blowholes lying side by side.

• **Rorquals** Rorquals are named for a Norwegian phrase meaning "furrow whale," referring to the pleats or grooves on the throat. Rorquals are gulpers; they take in large quantities of water when they feed. The pleats allow the throat to expand massively. The water is squeezed through the baleen, filtering out small fish or planktonic organisms.

• **Right whales** The three right whales—the bowhead, southern right whale, and northern right whale—form the family Balaenidae. They were named "right" by medieval whalers who considered them the best whales to catch; right whales swim slowly, migrate along regular coastal routes, and (because of their thick layer of blubber) float when dead. They have a very large head, with an upcurved upper jaw from which hang long baleen plates. Southern and northern right whales have bumps of rough skin, called callosities, on their heads. The bumps encourage the growth of barnacles.

▲ *A breaching gray whale. The loud splash provides one of the ways that these whales communicate over short distances.*

• **Gray whale** The gray whale is extinct in the North Atlantic and lives only in the North Pacific and Arctic. An adult has a mottled body. Rather than a dorsal fin, it has a series of humps, or crenulations, running along the lower back. The barnacle-encrusted head has two or four throat grooves. The upper jaw of the gray whale is shorter and thicker than that of other baleen whales; it is used to dig up food-rich sediment from the seabed. The baleen has stiff bristles for straining invertebrates from the water.

FEATURED SYSTEMS

EXTERNAL ANATOMY Gray whales are baleen whales with moderately streamlined bodies, flippers shaped like hydrofoils, and a powerful, horizontally flattened tail for swimming. *See pages 376–379.*

SKELETAL SYSTEM The backbone acts as an anchor for muscles that flex the body and flippers, and raise the tail up and down. Hind limb bones are still present, though tiny and largely without function. *See pages 380–382.*

MUSCULAR SYSTEM Large muscles above and below the vertebral column power vertical movements of the tail for locomotion. *See pages 383–384.*

NERVOUS SYSTEM The brain of a baleen whale has a large and highly folded cerebrum. The cerebrum's extensive surface area provides room for the vast number of nerve cell interconnections needed to process and interpret substantial amounts of sensory information. *See pages 385–388.*

CIRCULATORY AND RESPIRATORY SYSTEMS Both systems ensure that oxygen reaches vital organs during dives. *See pages 389–391.*

DIGESTIVE AND EXCRETORY SYSTEMS Gray whales and other baleen whales swallow small prey whole and in large quantities. Their three-chamber stomach digests food mechanically and then chemically. *See pages 392–393.*

REPRODUCTIVE SYSTEM Internal male sex organs and hidden mammary glands assist streamlining and ease of movement through water. *See pages 394–395.*

External anatomy

Whales use their flipper-shaped forelimbs for steering but have no hind limbs. Instead, the boneless tail flukes propel these animals through the water. The flippers and tail flukes act as hydrofoils. In cross section they are shaped like the wings of an aircraft. This shape generates a force by causing the pressure of water above the fluke to be lower than the pressure below. Over the course of a tail beat, the net effect of this force acts forward. The force, called thrust, drives the whale through the water. Unlike flying birds, which require considerable upthrust to remain aloft, whales are nearly weightless in water and usually need little upthrust to keep them swimming level.

The importance of streamlining

For a swimming animal, drag (the resistance to motion when an object passes through a fluid) is a major physical force to overcome. A streamlined body shape, like a torpedo, helps reduce drag. That is why baleen whales do not have external ears. Similarly, a baleen whale's external reproductive organs, when not in use, are tucked inside the abdomen, so improving streamlining. A whale's skin is smooth and almost hairless; baleen whale skin releases an oily substance that reduces drag further. Keeping the leading edge of appendages like flukes and flippers as narrow as possible also helps minimize drag. By swimming beneath the water rather than at the surface (where disturbance creates waves), whales can save a great deal of energy. A baleen whale's two blowholes, positioned on top of the head, allow the animal to breathe while barely breaking the surface.

Unlike most toothed whales, which are predators of fast-swimming prey, most baleen whales feed on slow-moving crustaceans and other small invertebrates. Baleen whales can easily maintain moderate speeds. Blue whales and fin whales can swim at 19 mph (30 km/h) when threatened; even the slow-swimming gray whale reaches 13 mph (21 km/h).

▼ *Features of the external anatomy of a gray whale. The body (as with most aquatic vertebrates) is torpedo-shaped. This gives the best possible ratio of propulsive force to drag, helping the animal save energy as it swims.*

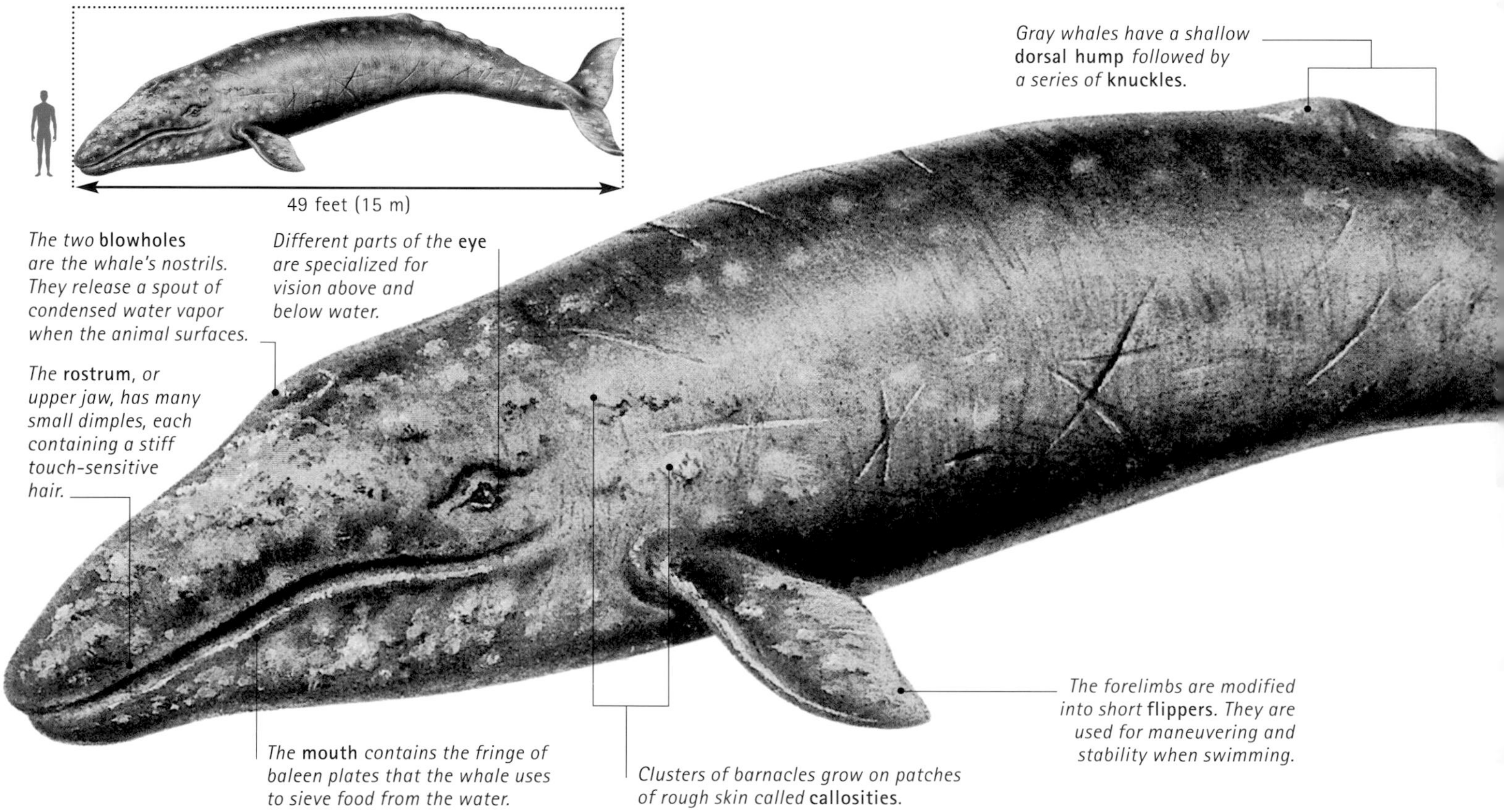

49 feet (15 m)
Gray whales have a shallow dorsal hump followed by a series of knuckles.
The two blowholes are the whale's nostrils. They release a spout of condensed water vapor when the animal surfaces.
Different parts of the eye are specialized for vision above and below water.
The rostrum, or upper jaw, has many small dimples, each containing a stiff touch-sensitive hair.
The forelimbs are modified into short flippers. They are used for maneuvering and stability when swimming.
The mouth contains the fringe of baleen plates that the whale uses to sieve food from the water.
Clusters of barnacles grow on patches of rough skin called callosities.

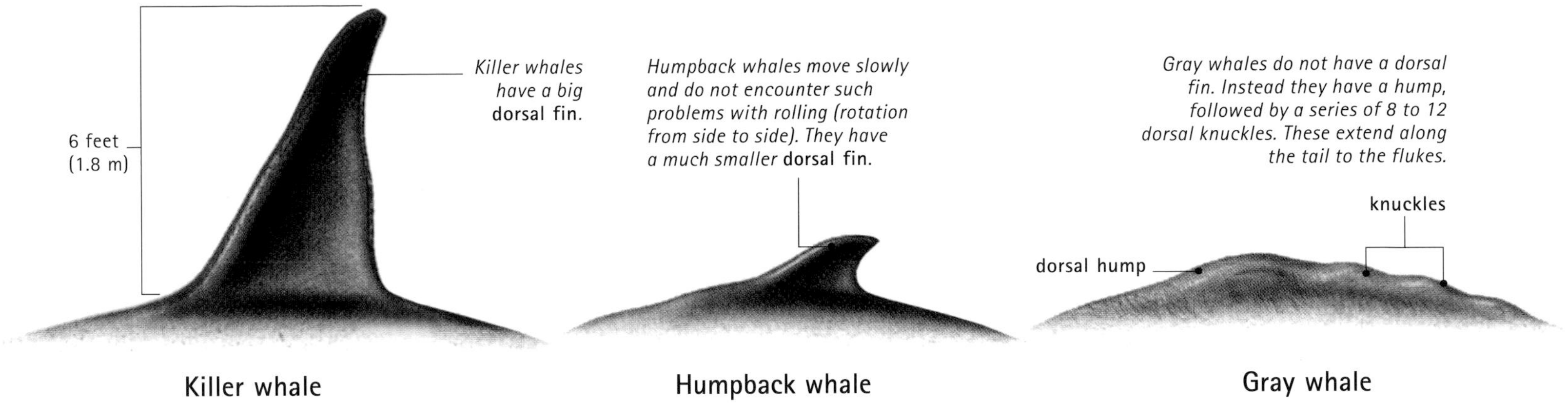

6 feet (1.8 m)
Killer whales have a big dorsal fin.
Humpback whales move slowly and do not encounter such problems with rolling (rotation from side to side). They have a much smaller dorsal fin.
Gray whales do not have a dorsal fin. Instead they have a hump, followed by a series of 8 to 12 dorsal knuckles. These extend along the tail to the flukes.
knuckles
dorsal hump
Killer whale
Humpback whale
Gray whale

Large size

Most baleen whales are very large animals. The blue whale may be the largest animal that has ever lived. It reaches lengths of up to 110 feet (33 m) and weighs up to 200 tons (180 metric tons). This is equivalent to the weight of more than 30 male African elephants, the largest living land animals, and is at least twice the size of the biggest known dinosaurs. Land animals cannot reach the sizes of these marine giants, since their limbs would need to be enormous to support their weight in air. Such limbs would not be strong enough for the job.

In cooler environments large size gives animals an advantage. Whales are warm-blooded, or endothermic: they control their body temperature physiologically rather than by relying on environmental or behavioral control. Because baleen whales have a core body temperature of 97–99°F (36–37°C), and because most of them live in water at temperatures at least 27°F (15°C) cooler than this for part of the year, reducing heat loss is of great survival value. For this reason, whales have thick layers of insulating blubber.

Large whales also have physics on their side. A small animal has a much higher surface-area-to-volume ratio than a large one. Therefore, large animals lose or gain heat from their surroundings more slowly, and need to use less energy to maintain their body temperature at a near-constant level.

▲ **DORSAL FINS**
Cetaceans
The dorsal fins of cetaceans help the animal stabilize as it swims. This is more important for faster-swimming species. Slower species usually have smaller fins, and gray whales have dispensed with them altogether. They instead have a row of smaller humps that run along the lower back.

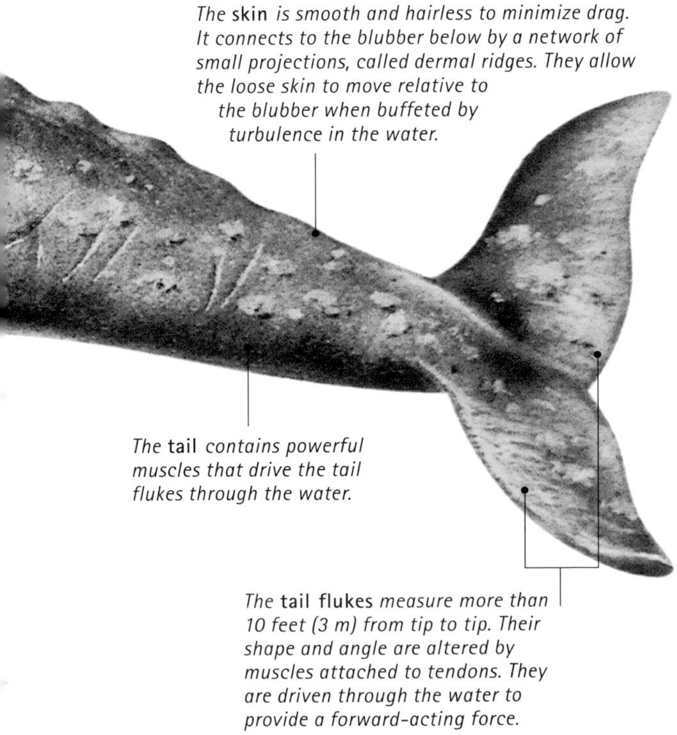

The skin is smooth and hairless to minimize drag. It connects to the blubber below by a network of small projections, called dermal ridges. They allow the loose skin to move relative to the blubber when buffeted by turbulence in the water.
The tail contains powerful muscles that drive the tail flukes through the water.
The tail flukes measure more than 10 feet (3 m) from tip to tip. Their shape and angle are altered by muscles attached to tendons. They are driven through the water to provide a forward-acting force.

EVOLUTION

Relatives of baleen whales

Baleen whales split from the toothed whales around 33 million years ago. The distribution of early mysticete fossils—along with abundant zooplankton fossils—suggests that the baleen whales evolved in the South Pacific and Antarctic before swiftly spreading throughout tropical and more northerly oceans. Baleen whales' toothed ancestry is betrayed by the presence of tooth buds in their developing embryos. These structures develop into teeth in toothed whales, such as sperm whales and dolphins. In baleen whales the tooth buds do not develop further, but baleen forms instead.

Studies of whale DNA (genetic makeup) and blood composition, together with anatomical evidence from living and fossil species, show that whales' closest relatives are the artiodactyl (even-toed) ungulates. Artiodactyls are hoofed mammals that include groups such as camels, sheep, and antelope. The closest living relatives of whales are probably hippopotamuses.

▲ *Whale lice shelter among barnacles on a gray whale. Swimming barnacle larvae are encouraged to settle on rough patches of skin around the whale's head called callosities. The barnacles then transform into sessile (attached) adults.*

Baleen

Baleen plates are made of keratin—the same tough, flexible protein that hair and fingernails are made of. Baleen plates have a consistency and springiness similar to that of human fingernails. Depending on the species, between 140 and 430 baleen plates hang in two rows from both sides of the upper jaw. The plates are arranged like rows of kitchen dishes stacked in a drying rack. As the tip of a baleen plate wears down, new keratin is added at the base of the plate where it is embedded in the gum.

Seen from the front, a baleen plate is triangular, with its inner edge a fringe of fibers rather like the splayed-out bristles of a broom. The size, flexibility, and number of baleen plates govern the foods that the whale eats.

Unwanted passengers

Patches of roughened skin on the body of gray, humpback, and right whales attract barnacles. The barnacles gain a safe home from which they can feed freely on small plankton floating in the water. Whale researchers use the pattern of barnacle growths as identification markers.

The purpose of these barnacle clumps, or callosities, is unclear. They reduce the whale's streamlining and seem to serve no obvious benefit. However, male right whales sometimes use their callosities as weapons in fights for females, raking their opponents with the abrasive patches. Patterns of callosities may also help whales identify one another.

Gray and right whales carry thousands of parasitic crustaceans called cyamids, or whale lice. They feed on flecks of discarded skin, finding a refuge around the barnacles. Some researchers have suggested that the callosities encourage barnacles and lice to concentrate only on certain parts of the body, so the rest is less affected and therefore maintains its streamlining. Whales are sometimes host to larger hitchhikers. Remora fish use suckers on their heads to attach to passing whales, which unwittingly give them a free ride.

Body coloration

The body of a baleen whale is typically a combination of gray, black, brown, or white. Humpbacks have distinctive patches of black and white on their flippers and the underside of their flukes. Scientists use the fluke markings to identify individuals. Fin whales have a pale patch on the right side of the head.

CONNECTIONS

COMPARE the behavior of whale lice with the ***LOUSE*** that parasitizes land mammals.

COMPARE the baleen plates of a gray whale with other filtering mechanisms, such as that of a ***GIANT CLAM***.

CLOSE-UP

Blubber

Like other marine mammals, whales have a thick fat-rich layer called blubber beneath the skin. Whale blubber is a mixture of fibrous connective tissue interspersed with oil-filled cells. Blubber insulates the whale against the cold, acts as a food store, and, being less dense than water, provides buoyancy. With the exception of Bryde's whale, all baleen whales inhabit cool temperate, subpolar, or polar waters for at least part of the year. The blubber layer of large baleen whales is at least 5 inches (12 cm) thick. The bowhead whale, which spends all year in subpolar or polar waters, has a blubber layer up to 20 inches (50 cm) thick.

The giant liver of baleen whales is also an important fat store, providing energy reserves during winter when food is scarce. The blubber of gray whales, together with fat stores in the bones and liver, sustains the adults during their seven-month round trip migration between polar and tropical waters.

Scientists think the whales use this patch to scare fish into a tight school at the water's surface. Fin whales typically turn on their right side when feeding, and sometimes they cooperate in a group to encircle their prey.

A baleen whale's upper surface is typically darker than its under surface. That is called countershading. The pale underside makes the silhouette of the body much less visible against the background of sunlight streaming through the surface water. This coloration may act as camouflage against predators attacking from below. The dark upper surface makes the whale less visible from above against the dark, inky depths of the sea. Countershading may be a legacy of the ancient past, when whales were hunted by a wider range of predators than they are now. Nowadays, adult baleen whales have few predators—only killer whales working as a team can kill an adult whale. Human whalers, however, are hunters against which the mightiest whales have no defense.

COMPARATIVE ANATOMY

Swimming underwater

Fish, whales, penguins, and ichthyosaurs (a group of extinct marine reptiles) are only distantly related to one another, but over millions of years of evolution all have adapted to swim through water. The development of a shape that allows optimal streamlining and the development of appendages such as flippers or fins are examples of convergent evolution. This occurs when distantly related organisms develop similar anatomical solutions to similar environmental demands. However, the propulsive tails of fish and ichthyosaurs move (or moved) from side to side, whereas those of whales move up and down. Penguins generate all the thrust they need to swim by flapping their flipperlike wings.

▼ A blue whale at its breeding grounds off Mexico. Note the pleats on the throat. These allow enormous expansion, enabling the whale to take a massive amount of water into its mouth.

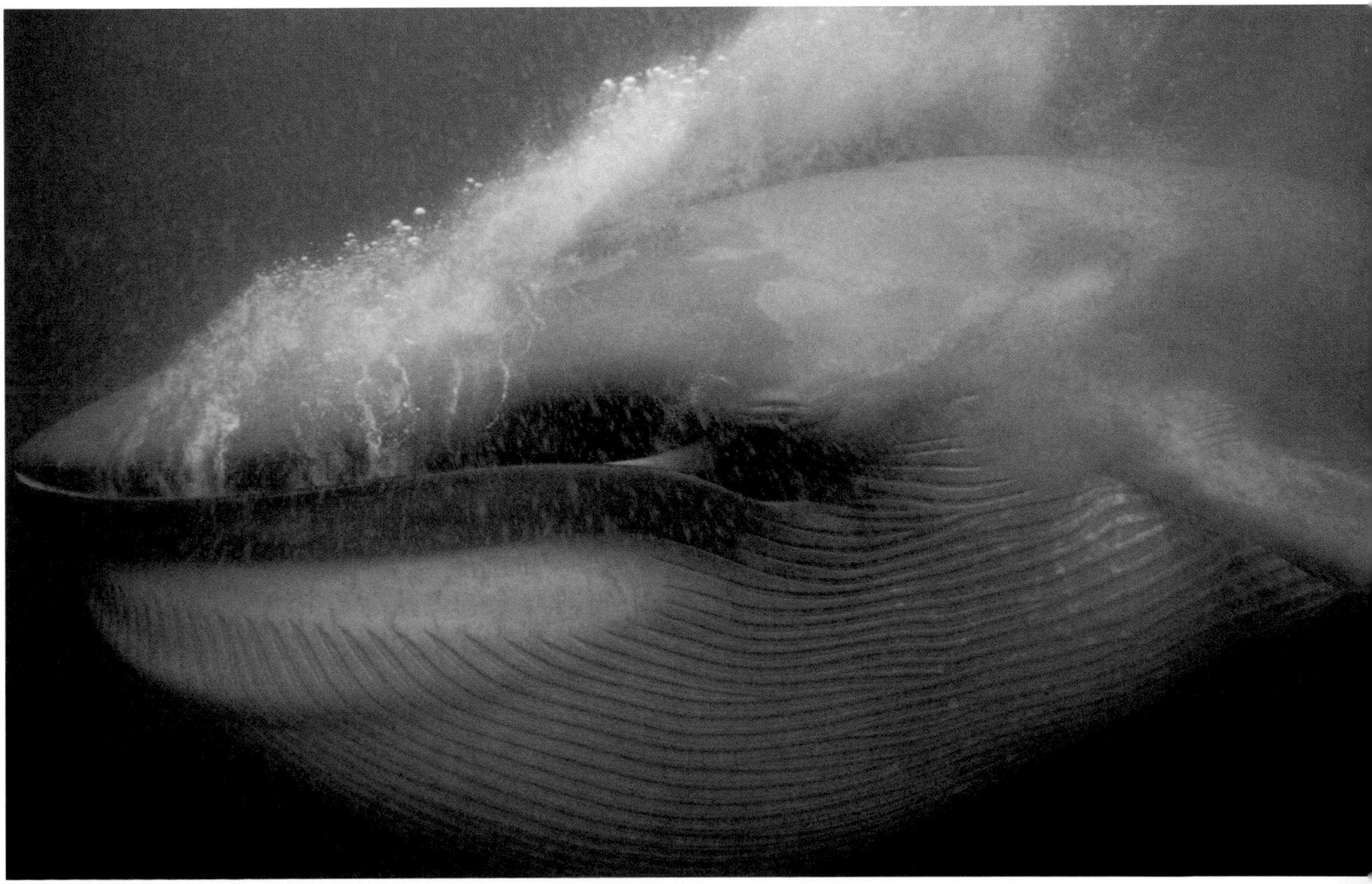

Skeletal system

CONNECTIONS

COMPARE the skeleton of a gray whale with that of a semiaquatic mammal, the ***HIPPOPOTAMUS***.

COMPARE the forelimb anatomy of swimming animals such as a gray whale and a ***PENGUIN***.

Vertebrate skeletons have three main functions: to support the animal's body; to protect vital internal organs such as the brain, heart, and lungs; and to allow movement of body parts such as the head and limbs, enabling locomotion. A baleen whale's skeleton is very different from that of a terrestrial (land-living) vertebrate because a marine mammal is buoyed up and supported by the water around it. For a terrestrial animal, such as a zebra, air provides little physical support. So terrestrial mammals need a strong skeleton with limbs that raise the body off the ground for efficient locomotion. Their limbs act as vertical compression struts, rather like the piles that support a bridge.

Mammals that spend their entire lives in water do not need this kind of support. A baleen whale with its lungs filled with air is effectively weightless in water. However, when it becomes stranded on the shore, its skeleton cannot support its body weight and the animal can be suffocated by its own weight.

Skull and jaws

The skull of a baleen whale is stretched lengthwise compared with that of a horse. The jaws are elongated, and the upper jaw, or maxilla, is fused to the skull. This fusion provides the strength to support the baleen plates that hang down from the maxilla.

Over millions of years of evolution, the nasal passages have moved to the top of the skull. Baleen whales breathe through two blowholes. Their position ensures that the whale can breathe without having to break the surface with all of its body. This helps the animal save energy as it swims.

The backbone

The backbone of a land mammals acts as a firm girder that supports the animal's weight in air. Water supports a whale's weight, so its backbone becomes relatively more important for locomotion than for support. Water is a dense, relatively viscous (thick) fluid, and is

▼ *The skeletal system of a gray whale. The hyoids (not shown) are small, delicate bones just beneath the skull. They are very important, since they serve as an attachment for the animal's massive, 1-ton tongue. The tongue is used to force water and sediment containing food through the baleen plates.*

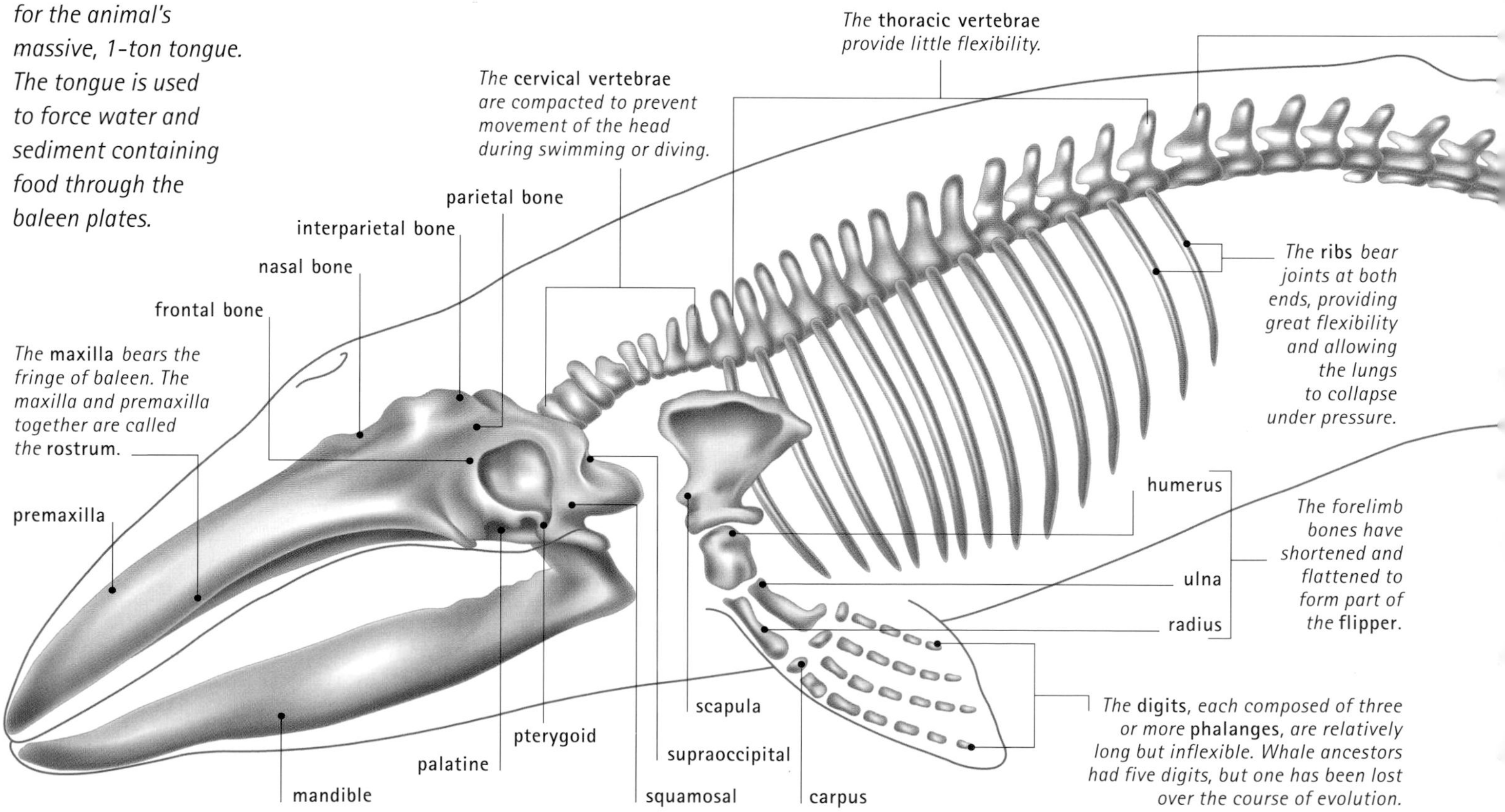

The thoracic vertebrae provide little flexibility.
The cervical vertebrae are compacted to prevent movement of the head during swimming or diving.
parietal bone
interparietal bone
nasal bone
frontal bone
The maxilla bears the fringe of baleen. The maxilla and premaxilla together are called the rostrum.
premaxilla
The ribs bear joints at both ends, providing great flexibility and allowing the lungs to collapse under pressure.
humerus
ulna
radius
The forelimb bones have shortened and flattened to form part of the flipper.
The digits, each composed of three or more phalanges, are relatively long but inflexible. Whale ancestors had five digits, but one has been lost over the course of evolution.
scapula
supraoccipital
pterygoid
palatine
mandible
squamosal
carpus

COMPARATIVE ANATOMY

Skulls of baleen whales

The skull and jawbones of baleen whales are remarkably distorted compared with those of all other mammals, including toothed whales. Over the course of evolution, plates at the back of the skull called the supraoccipital bones have shifted forward; other cranial bones have merged with those of the upper jaw to provide support for the baleen plates. The upper and lower jaws are extremely elongated, producing the wide gape that the whale needs to engulf vast quantities of food-rich water. Structure is closely linked to feeding strategy. In rorquals such as the humpback whale, the mouth cavity is broad and the rostrum is slightly curved, accommodating short baleen plates. In the gray whale, the upper and lower jaws are thickened, enabling the whale to "plow" the seabed for crustaceans, which are strained through shorter baleen plates. Right whales have wider jaws, with the rostrum highly arched and supporting long baleen plates. This feature is most extreme in the bowhead whale.

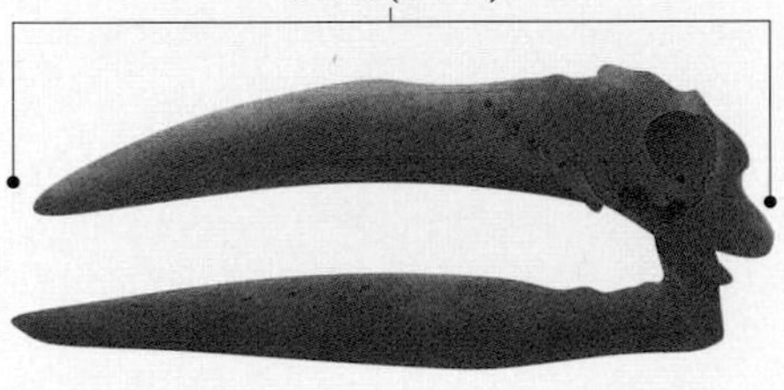

Gray whale
Short, robust mouthparts withstand the force of impact as the animal plows the seabed to stir up invertebrates, which are subsequently sieved from the water.

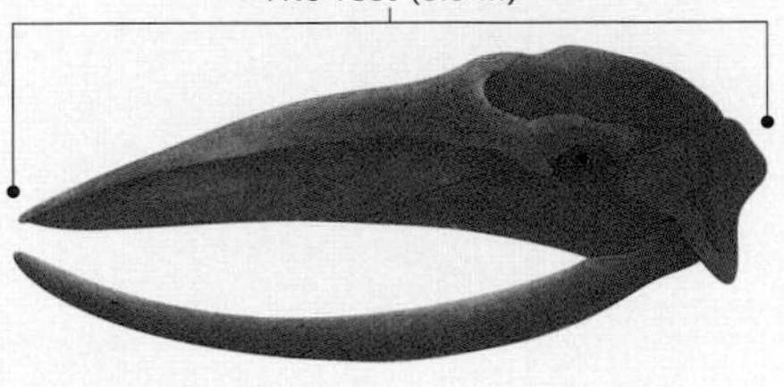

Humpback whale
The rostrum is more slender, longer, and gently curved. Humpbacks are gulpers that alternately swim, then gulp a mouthful of plankton or small fish.

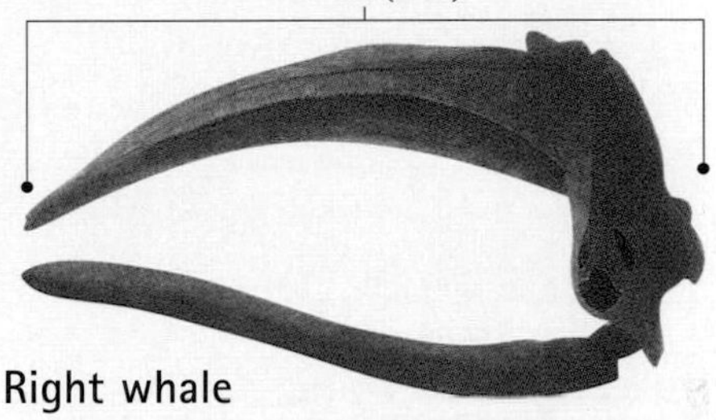

Right whale
With a hugely arched, elongated, and slender upper jaw, right whales are skimmers; they keep their mouths open all of the time to filter plankton.

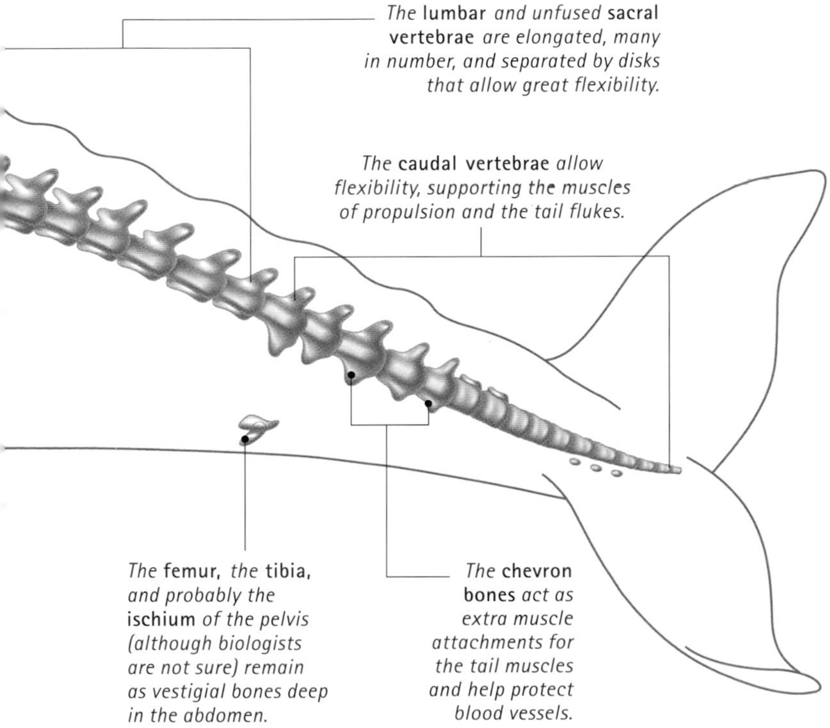

The lumbar and unfused sacral vertebrae are elongated, many in number, and separated by disks that allow great flexibility.
The caudal vertebrae allow flexibility, supporting the muscles of propulsion and the tail flukes.
The femur, the tibia, and probably the ischium of the pelvis (although biologists are not sure) remain as vestigial bones deep in the abdomen.
The chevron bones act as extra muscle attachments for the tail muscles and help protect blood vessels.

difficult to shift. A whale's locomotory muscles need to be large and well anchored. In the blue whale, muscles account for about 40 percent of the animal's weight, and the skeleton for only about 17 percent.

In most land mammals, the backbone flexes most somewhere between the last part of the thoracic (chest) region and the lower back (lumbar) region. Limbs close to these regions are responsible for the forces that move the animal from one place to another. In whales, however, it is the tail that propels the animal forward. The regions of greatest flexibility in these animals are at the base of the tail and around the tail flukes. Sections of the vertebral column in front of the tail are fairly rigid.

A gray whale typically has 56 vertebrae. In baleen whales, the neck is short, and the seven cervical vertebrae are flattened along the axis of the spine. The neck moves little. This ensures that side-to-side movement of the head, which would increase drag and instability, does not occur while the whale is swimming. The number of thoracic vertebrae is about the same (14) in whales and land mammals, but those of whales permit much less movement. The ribs attached to them, however, are very mobile. They allow the area enclosed by the rib cage to alter greatly in volume. This enables the lungs to collapse during dives. The lumbar region is elongated in whales in comparison with land mammals. Either the lumbar vertebrae are stretched from front to back, or there may be more of them. Each vertebra is large, with broad surfaces and extensions called neural spines and transverse processes. They act as attachment points for the tail muscles. The neural spines act as levers, increasing the power that can be transmitted by the muscles to make the spine flex.

The first caudal vertebra is level with the anus. There are usually more caudal vertebrae in a whale (up to 23) than in most land mammals. The intervertebral joints—each containing a cushioning pad of cartilage called a disk—give suppleness and elasticity to the tail. Most caudal vertebrae have large neural spines and transverse processes for muscle attachment. Pairs of chevron bones are present on the ventral (belly) side of the caudal vertebrae, except those that support the flukes. Chevron bones enclose and protect major blood vessels from damage when the tail flexes. The last few caudal vertebrae are flattened bones that support the tail flukes.

Limbs and their supports

In running quadrupedal land mammals, the forelimbs act mainly as shock absorbers, while the hind limbs provide thrust. The limbs connect to the spine through limb girdles. The front (pectoral) girdle typically contains two scapulae (shoulder blades) and clavicles (collarbones). Over millions of years, whales have lost their clavicles, and the front limbs of whale's ancestors evolved into flippers. Compared with a human arm, the skeleton of a flipper is similar but with shorter upper and lower arm components and longer digits. The major joints in the flipper are inflexible but are not completely fused.

In a land mammal, the rear, or pelvic, girdle is more robust than the shoulder girdle. The pelvic girdle is anchored to the sacral vertebrae of the spine, which are fused. During whales' evolutionary transition over millions of years from terrestrial to fully aquatic life, their sacral vertebrae have become unfused. This allows enhanced mobility and flexibility of the tail. The hind limbs have shrunk dramatically in size, along with the pelvis.

The buds of hind limbs are present in whale embryos, but do not develop fully. Traces of the tiny hind limbs and part of the pelvis remain inside adult baleen whales, embedded deep within the muscle of the abdomen. These little bones are vestigial structures—they serve no function at all.

CLOSE-UP

The densest of bones

The long bones of whales are light. They do not need to be strong enough to support the animal in air. They consist mostly of spongy bone with a thin outer shell of harder, more compact bone. However, things are very different in the rarely seen Blainville's beaked whale. The rostrum of this species is extremely dense, almost as dense as a tooth; a Blainville's beaked whale rostrum contains the densest bone known to science. Why do these whales need to deposit such dense bone around the snout, a process that uses significant amounts of energy?

Superdense bone may help the whales dive or transmit sounds in some way. However, biologists have recently shown that the rostrum becomes secondarily ossified (extra bone is deposited) only when the whales become sexually mature. This suggests that the bone may be used as a shield to prevent damage during aggressive encounters between males. They may thrash their heads against one another in battles for dominance. However, due to the elusiveness of this species, the details of its reproductive behavior remain completely unknown.

▼ **THORACIC VERTEBRA**
Baleen whale
Thoracic vertebrae are located at the trunk.

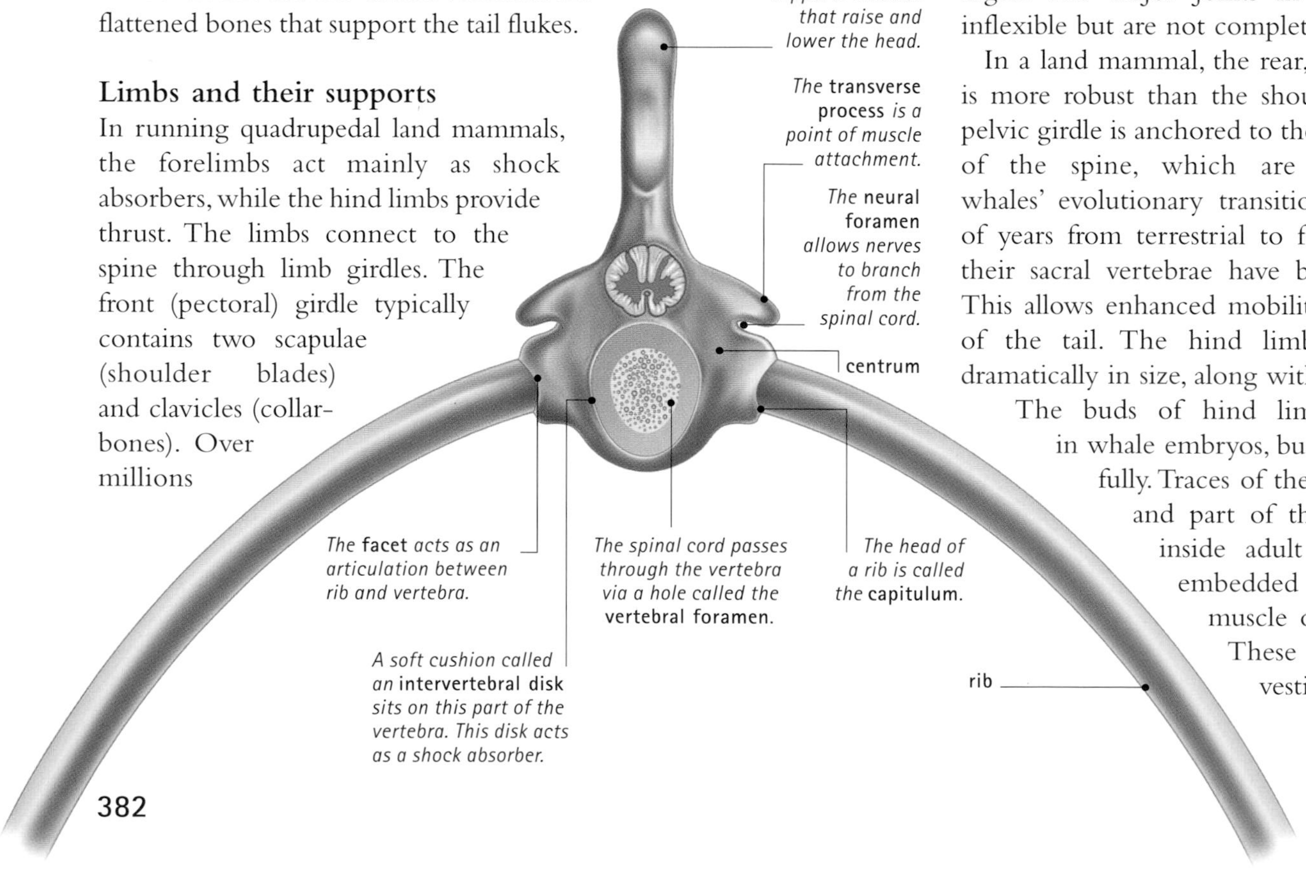

The neural spine supports muscles that raise and lower the head.
The transverse process is a point of muscle attachment.
The neural foramen allows nerves to branch from the spinal cord.
centrum
The facet acts as an articulation between rib and vertebra.
The spinal cord passes through the vertebra via a hole called the vertebral foramen.
The head of a rib is called the capitulum.
A soft cushion called an intervertebral disk sits on this part of the vertebra. This disk acts as a shock absorber.
rib

Muscular system

Over millions of years of evolution, many of the head muscles in whales have moved to different locations from those of their terrestrial relatives. Some of these muscles have also taken on new functions. For example, some head muscles are involved in closing the blowholes to prevent water from entering when the whale dives. In baleen whales, muscles extending between the lower jaws squeeze water through the baleen.

A large sheet of muscle, called the cutaneous trunci, lies beneath the skin and fat layer and covers most of the thorax and abdomen in many mammals. In cetaceans, part of this muscle is modified for squeezing milk out of the female's mammary gland and into the mouth of nursing young.

Limb, trunks, and tails

The forelimbs of a whale have limited movement compared with those of most land mammals. Movement of a whale's flipper is largely limited to upward or downward flexes that aid steering and control. The number and size of muscles are reduced overall, particularly toward the lower end of the limb. Muscles between the bones give the flipper some flexibility. Many whale trunk and tail muscles are unusual when compared with those of land mammals. The muscles wrap around the vertebral column and rib cage like a sheath. This makes it difficult to distinguish one set of muscles from another. Many of the muscles are elongated along the axis of the vertebral column; these muscles contract over long distances.

COMPARATIVE ANATOMY

A movable penis

Male whales have a muscle that withdraws the penis into a pouch within the body. Artiodactyl mammals also have this muscle, but it is absent in other mammals. In cetaceans, the rectus muscle—together with alterations in blood flow—enables the male to move his penis. In addition to mating, male whales use this prehensile penis as a sensory appendage in social encounters.

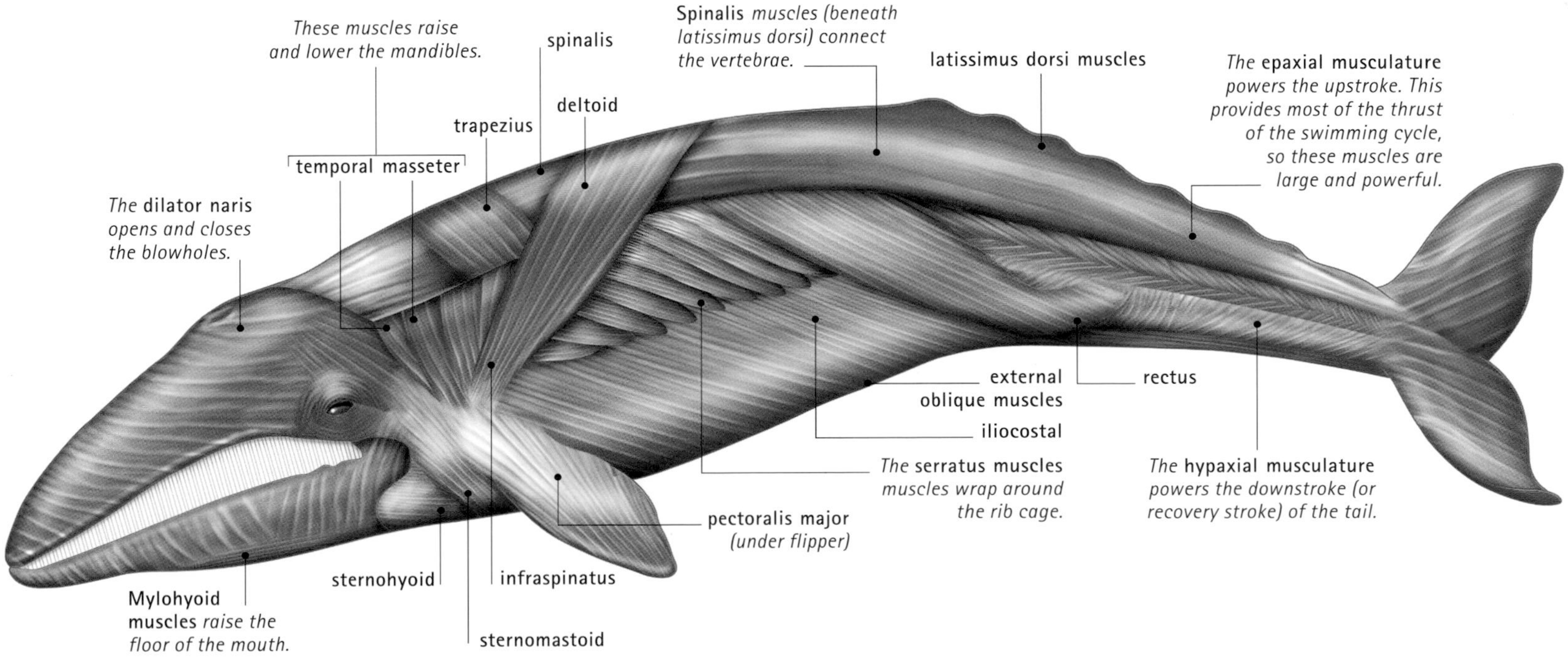

These muscles raise and lower the mandibles.
spinalis
Spinalis muscles (beneath latissimus dorsi) connect the vertebrae.
latissimus dorsi muscles
The epaxial musculature powers the upstroke. This provides most of the thrust of the swimming cycle, so these muscles are large and powerful.
deltoid
trapezius
temporal masseter
The dilator naris opens and closes the blowholes.
external oblique muscles
rectus
iliocostal
The serratus muscles muscles wrap around the rib cage.
The hypaxial musculature powers the downstroke (or recovery stroke) of the tail.
pectoralis major (under flipper)
sternohyoid
infraspinatus
Mylohyoid muscles raise the floor of the mouth.
sternomastoid

▼ *Features of a gray whale's muscular system. The mylohyoid muscles are particularly important for baleen whales. They raise the floor of the mouth, forcing water through the baleen plates.*

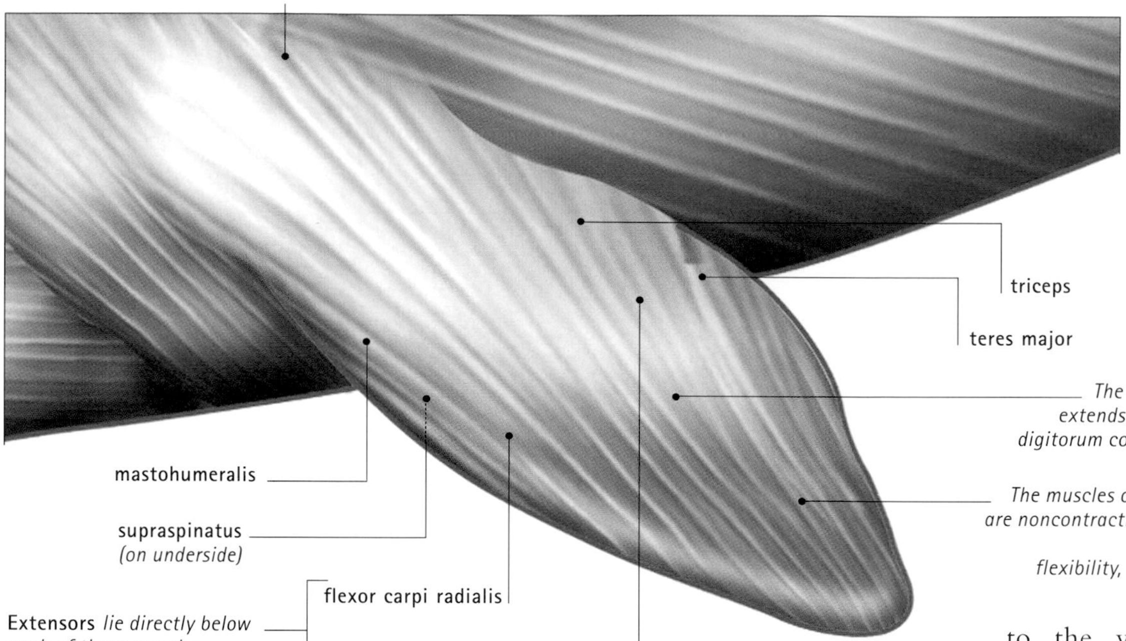
deltoid
triceps
teres major
The extensor digitorum communis extends the digits, or fingers. The flexor digitorum communis lies directly beneath it.
The muscles of the fingers are much reduced, are noncontractile, and act more like ligaments. They provide some elastic flexibility, helping the animal save energy.
mastohumeralis
supraspinatus (on underside)
flexor carpi radialis
Extensors lie directly below each of these muscles.
flexor carpi ulnaris

◀ **FLIPPER MUSCLES**
The forelimb muscles of cetaceans are much reduced when compared with those of land mammals, particularly at the distal (lower) end.

Powering swimming

Expansion and contraction of the muscles attached to the tail enable it to bend up or down to power swimming. Lying just above the vertebral column, the epaxial muscles connect to the projections, or transverse processes, of the lumbar and caudal vertebrae. The epaxial muscles extend back to other vertebrae, inserting into a sheath of tendons connected to the vertebral column that distributes power widely and smoothly along the tail. When these muscles contract, they bend the tail upward. This provides the bulk of the thrust required for forward swimming.

The hypaxial muscles lie beneath the vertebral column. They attach to the transverse processes of the thoracic and lumbar vertebrae; at the other end they are inserted into a sheet of tendons connected to the caudal vertebrae and chevron bones in the tail. These muscles contract to power the tail's recovery stroke, when it moves downward prior to another powerful upstroke.

▼ *The caudal muscles control the angle of the flukes of this humpback whale's tail.*

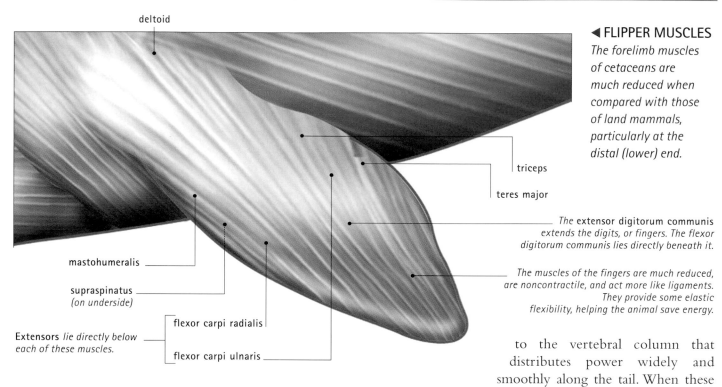

Nervous system

Like primates, whales have unusually large and complicated brains, but a baleen whale's brain is smaller, relative to body mass, than that of a toothed whale or a human. Both whales and primates have a large, folded cerebrum region containing a pair of cerebral hemispheres. The cerebrum is responsible for learning, reasoning, and memory, as well as the processing of complex sensory information, such as that relayed from the eyes and ears.

In baleen whales, more than 68 percent of the brain's weight is accounted for by the cerebrum; in humans the figure is 83 percent or more. Various parts of a human's cerebrum appear different, and control and coordinate different functions. In whales, different parts tend to have similar structures even when they control different functions. According to one theory, the parts of a whale's cerebrum look the same because its brain tissue has to develop more rapidly than that of primates. By the time a whale calf is born, it has to be able to swim. Humans, by comparison, are helpless at birth and cannot walk until they are at least a year old. Human brains develop more slowly, and therefore their tissues have much longer to differentiate.

IN FOCUS

Magnetic sense

Scientists have tracked fin whales swimming along lines of equal magnetic force as they migrate from one region to another. Many baleen whales have tiny crystals of magnetite (an iron oxide) in parts of the brain. The crystals act like miniature magnetic compasses. This suggests that some whales at least are able to navigate using Earth's magnetic field. This ability might help explain how some baleen whale species can navigate across vast distances through the oceans when they do not have access to landmarks. Some birds have a similar magnetic sense.

CONNECTIONS

COMPARE the size of the brain of a gray whale, a ***HUMAN***, and a ***DOLPHIN***. In absolute terms a gray whale has a massive brain, but it is actually only a little larger than expected for an mammal of such a size.

The cerebellum

The second largest region of a whale's brain is called the cerebellum. It is concerned with relaying and coordinating the position of the whale's body in space and time. The cerebellum continually adjusts body movements and relays information to the

▼ *The gray whale's nervous system is similar to that of other cetaceans. Note the cranial nerves. Three are shown here, but whales and other mammals have 12 in total. These may have sensory or motor functions, or a combination of both.*

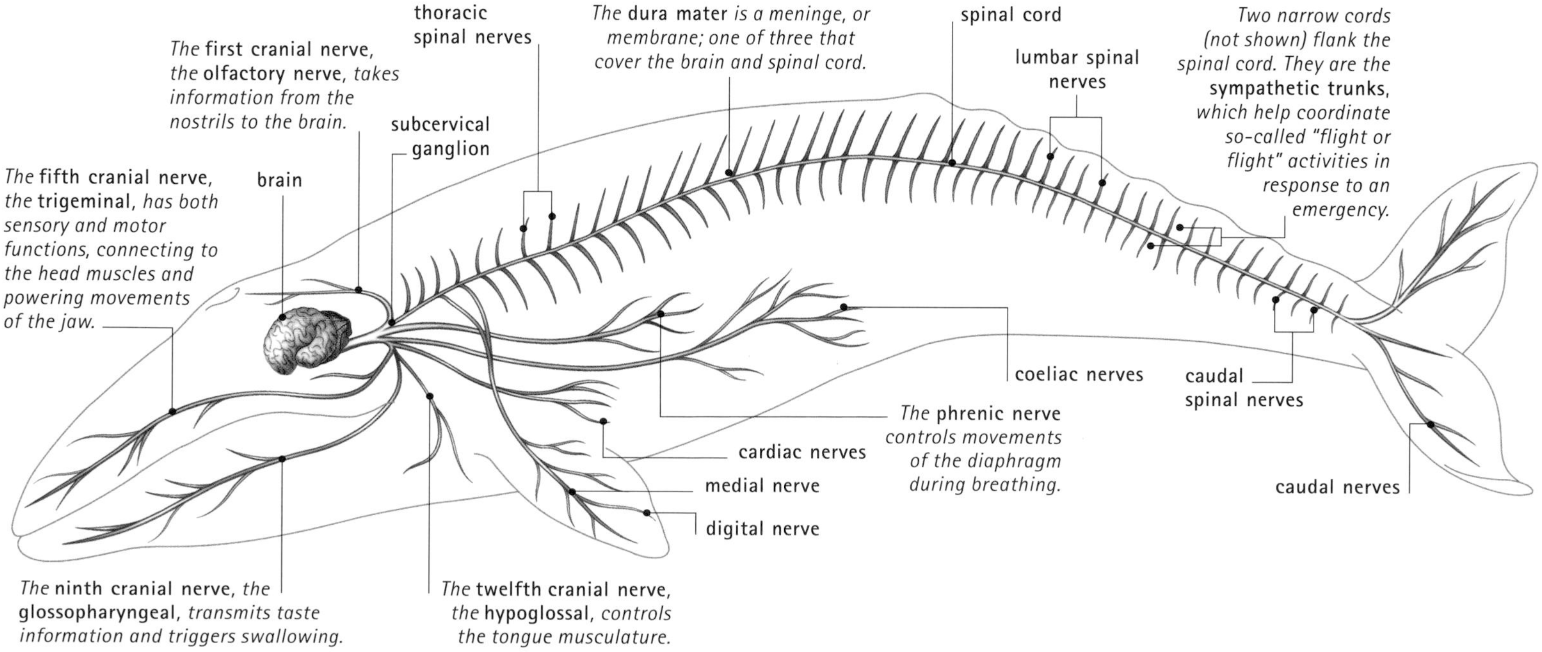

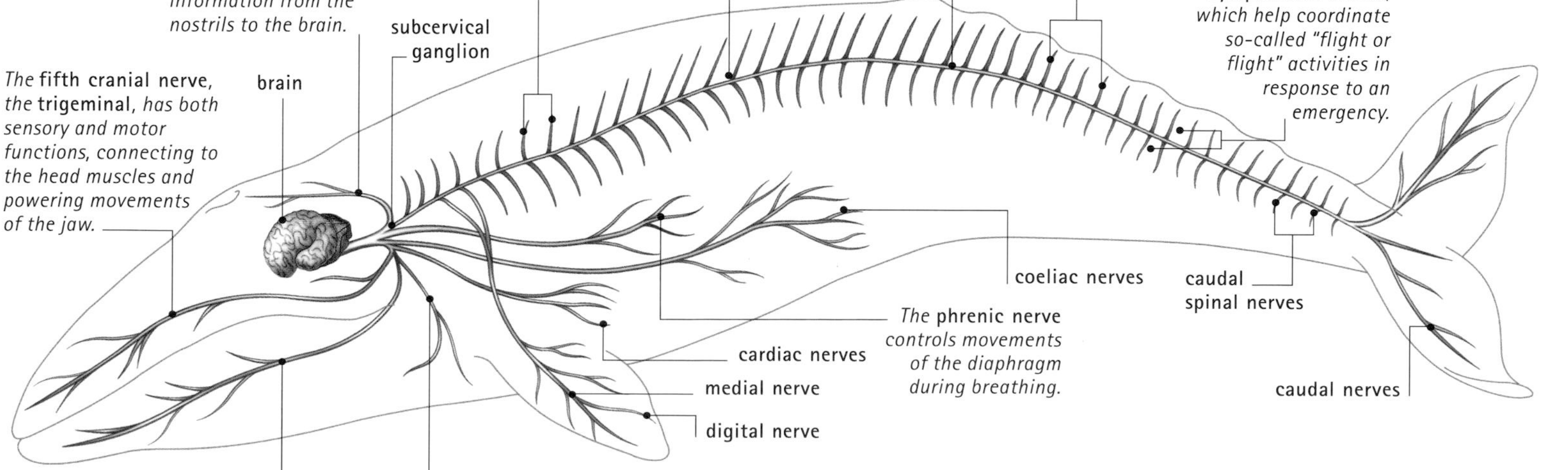

thoracic spinal nerves
The first cranial nerve, the olfactory nerve, takes information from the nostrils to the brain.
subcervical ganglion
The dura mater is a meninge, or membrane; one of three that cover the brain and spinal cord.
spinal cord
lumbar spinal nerves
Two narrow cords (not shown) flank the spinal cord. They are the sympathetic trunks, which help coordinate so-called "flight or flight" activities in response to an emergency.
The fifth cranial nerve, the trigeminal, has both sensory and motor functions, connecting to the head muscles and powering movements of the jaw.
brain
coeliac nerves
caudal spinal nerves
The phrenic nerve controls movements of the diaphragm during breathing.
cardiac nerves
medial nerve
digital nerve
caudal nerves
The ninth cranial nerve, the glossopharyngeal, transmits taste information and triggers swallowing.
The twelfth cranial nerve, the hypoglossal, controls the tongue musculature.

higher regions of the brain, including the cerebrum, for planning and decision making. The cerebellum of baleen whales accounts for at least 18 percent of the brain's weight; in humans it is only 10 percent. The relatively large size of a baleen whale's cerebellum reflects the fine degree of control that a swimming whale needs, requiring continual adjustments to its body movements.

Touch, taste, and smell

Baleen whales may be able to taste and smell using sensory receptors located in their mouth and nostrils. The tasting of food items, which in many cases are swarms of krill or shoals of fish, may be less important than in toothed whales, which select their prey individually. However, some experts believe that baleen whales can smell chemicals released by krill and other kinds of zooplankton. This ability may enable whales to seek out distant prey. Tasting seawater could also be important for detecting the presence of prey in the vicinity. It may also allow whales to track chemicals called pheromones, which indicate that a whale is ready to breed.

Touch is important in the lives of baleen whales. They sometimes stroke or touch one another with their flippers or other sensitive parts of their body. These acts are expressions of social bonding. Baleen whales have a scattering of tiny touch-sensitive bristles in the head region. Those of gray whales lie in small depressions around the head and jaws. The hairs help detect water flow around the head, and may also be important for detecting low-frequency sounds. As in toothed whales, the region around the blowhole is particularly sensitive. It is important for the whale to detect when the blowhole breaches the water surface. This triggers a breathing response. The sensitivity of this region also helps prevent potentially life-threatening damage to the blowhole that could impair breathing.

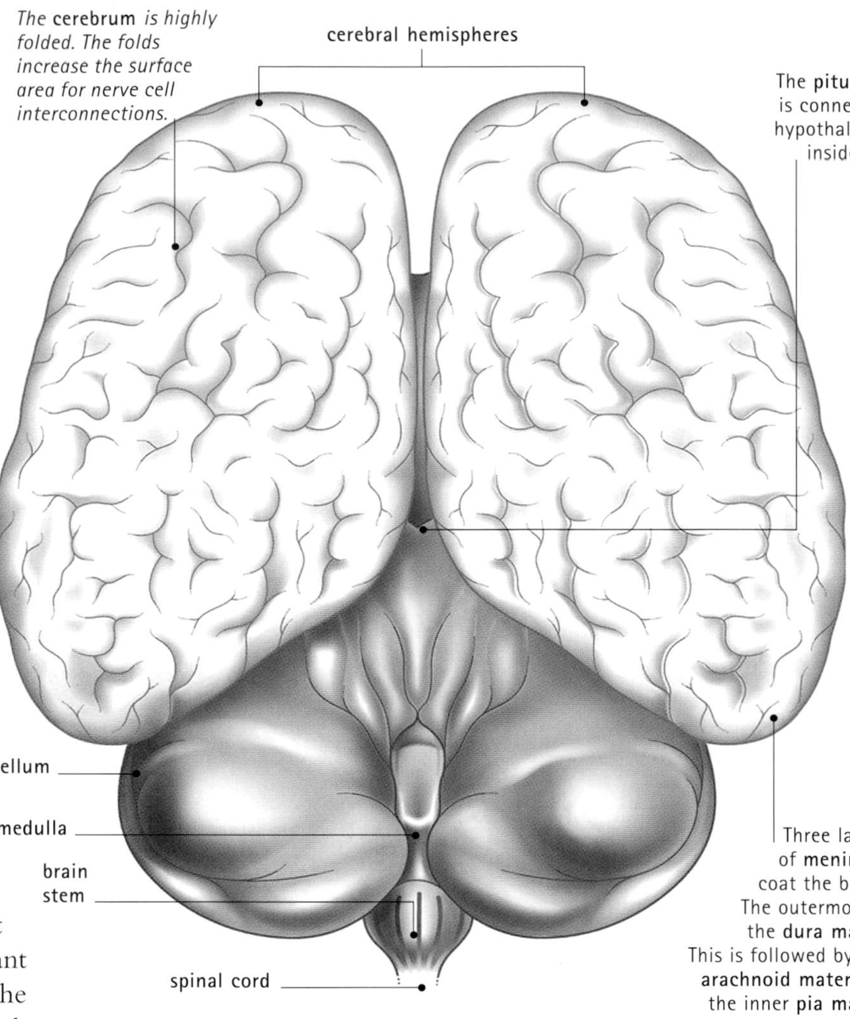

The cerebrum is highly folded. The folds increase the surface area for nerve cell interconnections.
cerebral hemispheres
The pituitary gland is connected to the hypothalamus, deep inside the brain.
cerebellum
medulla
brain stem
spinal cord
Three layers of meninges coat the brain. The outermost is the dura mater. This is followed by the arachnoid mater and the inner pia mater.

▲ BRAIN
Gray whale
A ventral (from beneath) view of the brain. The cranial nerves (not shown) depart from the underside of the brain. Most of these branch off from the medulla.

Whales' sight

Whales use vision to navigate around undersea obstacles, keep track of members of the social group, find prey at close quarters, and look out for potential predators. However, about 90 percent of sunlight is absorbed and reflected by a depth of only 30 feet (9 m). Whales must, therefore, be able to see in very poor light.

IN FOCUS

Spyhopping

Many baleen whales spyhop. They slowly rise out of the water until the head and eyes are above the surface. They look around, perhaps checking for coastline features or for the presence of other whales at the surface. A small social group (pod) of gray whales will often rises to the surface together.

However, they also need to be able to see in the bright light of the surface. The ability to see at extremes of light intensity, and both underwater and in air, is achieved in two ways. First, the pupil (the central hole that appears black) in each eye is able to alter drastically in size to adjust the amount of light entering the eye. The pupil is wide in dim light, but shrinks down to a small dot in the bright light of surface waters. A layer of light-detecting cells, the retina, is extremely sensitive, but the reduction of the pupil's diameter keeps it from being damaged.

Second, the eye lens is highly elastic and is spherical, unlike the flattened lenses of land mammals. Baleen whales see well in shallow water and moderately well in air. A reflective layer of cells lies behind the retina. When the whale is underwater, the reflective layer bounces light back through the retina, giving it a second chance to be absorbed. This helps whales see well in the dim light of deep water. Almost uniquely among mammals, whales can move their eyes independently of one another. Although baleen whales' eyes are on the sides of the head, the left and right fields of vision overlap slightly at the front. Thus, there is an area of binocular vision around the snout where the whale can see in three dimensions and so can judge distance accurately.

IN FOCUS

Whale songs

During the breeding season, adult male humpbacks float nearly vertical in the water, with their head uppermost, and start to sing. Their haunting songs, containing hundreds of notes ranging from rumbles and moans to squeaks, squawks, and chirps, advertise their presence to other humpbacks in the area. Whales that live in a certain area produce songs that are similar and share the same "dialect." The songs change from season to season, and from one locality to another.

Whales can see well only over relatively short distances. Light is more strongly refracted (bent) at the eye's surface when an eye is in air than when it is in water. As a result, an eye that can see well in air would be out of focus underwater. Gray whales avoid this problem by using different regions of the eye for focusing in air and in water. The outer layer of the eyeball is unusually thick to resist abrasion. This resistance is supported by the shedding of an oily protective fluid, rather than tears, onto the surface of the eye.

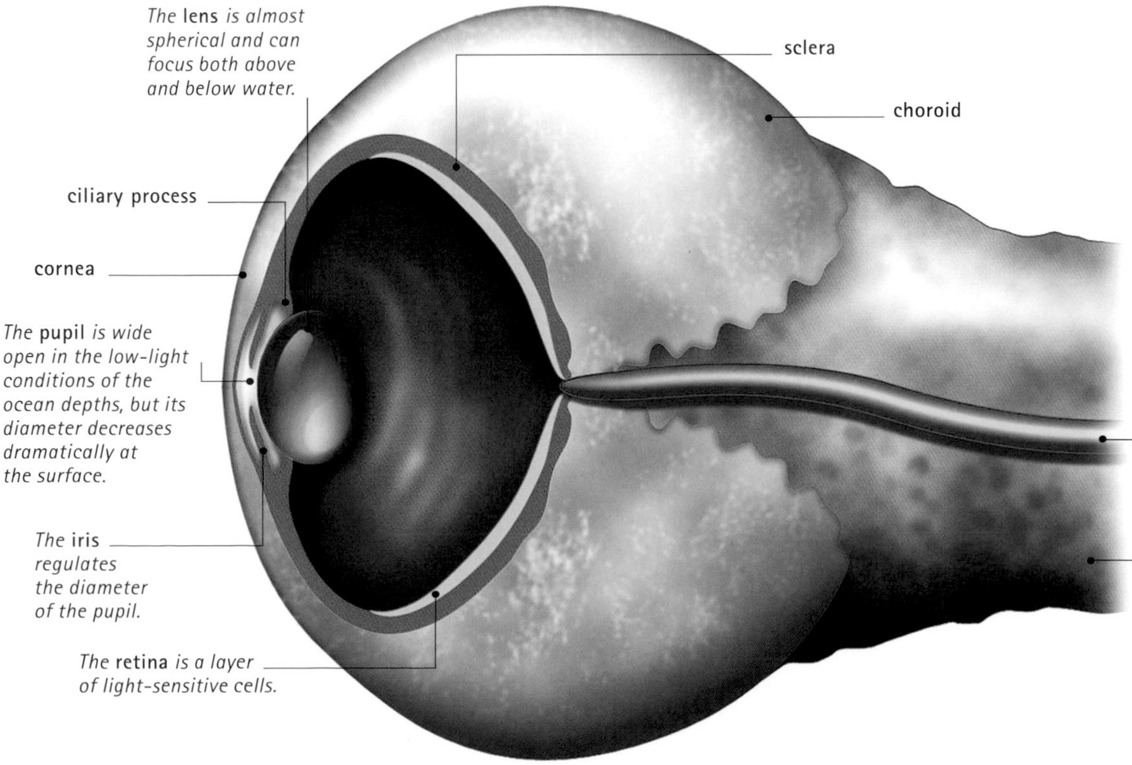

The lens is almost spherical and can focus both above and below water.
sclera
choroid
ciliary process
cornea
The pupil is wide open in the low-light conditions of the ocean depths, but its diameter decreases dramatically at the surface.
The iris regulates the diameter of the pupil.
The retina is a layer of light-sensitive cells.
The second cranial nerve, or optic nerve, transmits messages to the brain.
Vascular tissue supplies the eye with blood.

◀ **THE EYE**
Blue whale
The structure of a typical baleen whale eyeball. The choroid is extra-thick to resist wear. Note the spherical lens, and compare it with the lenses of land mammals, which are generally flatter.

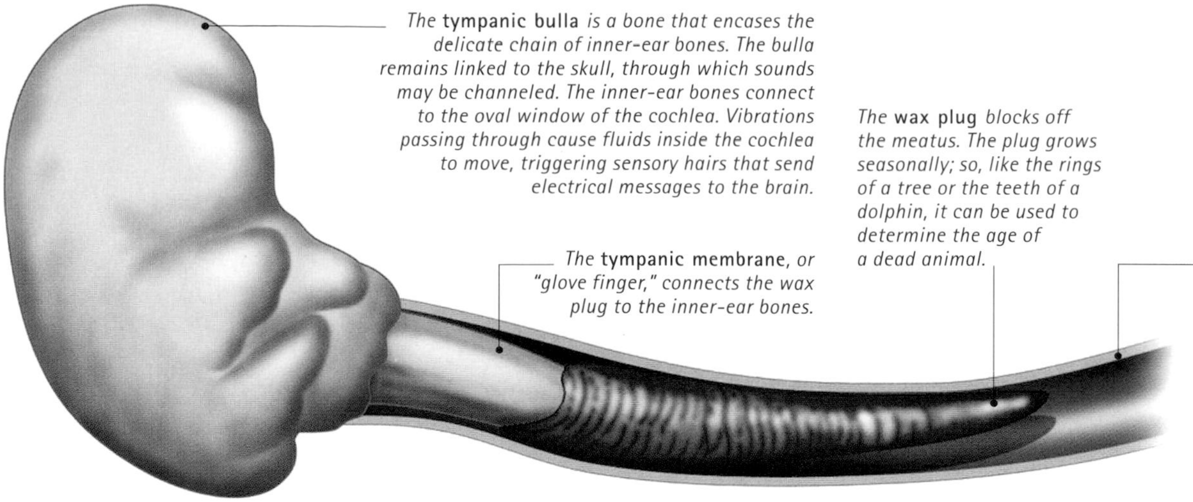

The tympanic bulla is a bone that encases the delicate chain of inner-ear bones. The bulla remains linked to the skull, through which sounds may be channeled. The inner-ear bones connect to the oval window of the cochlea. Vibrations passing through cause fluids inside the cochlea to move, triggering sensory hairs that send electrical messages to the brain.
The wax plug blocks off the meatus. The plug grows seasonally; so, like the rings of a tree or the teeth of a dolphin, it can be used to determine the age of a dead animal.
The tympanic membrane, or "glove finger," connects the wax plug to the inner-ear bones.
This tube, the external auditory meatus, connects the bulla to the outside world. The path of sound to the inner-ear bones is not well understood in baleen whales; this canal may be vestigial and serve no purpose.

▲ EAR STRUCTURE
Fin whale
A cross section through the head to show the bulla and meatus. How sound gets to the bulla is unknown. The auditory meatus may be important for airborne sounds, but biologists do not know for sure.

How whales hear

Sound travels farther and faster in water than in air, and baleen whales have good hearing. Sounds are detected and transmitted to the brain by means of a structure called the cochlea. Just how sound waves reach this structure is not clear. Sounds do not pass through the lower jaw, as in toothed whales. The outer ear opening in a baleen whale is tiny; it lies just behind the eye. The tube into which the ear opening leads, the external auditory meatus, is partly blocked by connective tissue and fully blocked by a waxy ear plug. This connects to a thick tympanic membrane, or eardrum. As in toothed whales, the bones of the inner ear are housed in a bony capsule called a tympanic bulla. Baleen whale bullae are not completely detached from the skull, suggesting that vibrations passing through bone and other tissues may be important. Air spaces in the skull called sinuses may help reflect vibrations through the skull bones toward the cochlea.

Regardless of how the vibrations get there, sounds cause vibrations of part of the cochlea called the oval window. These vibrations generate movements of fluids inside the cochlea. The fluids move sensory hairs, which send electrical signals to the brain.

Sound production

Baleen whales hear best at frequencies similar to the sounds they produce. Most make deep infrasonic sounds that lie outside the range of human hearing, as well as a range of moans, groans, whistles, and clicks that humans can hear. Bowhead and humpback whales are the most vocal, with male humpbacks producing complex songs that last for 10 minutes or more. Like toothed whales, humpbacks are also sensitive to high-frequency ultrasound, and other baleen whales probably are, too.

Hearing and sound production are vital for communication over both short and long distances. Baleen whales also use hearing to detect the sounds of their prey or of other predators, such as seals or seabirds, that make loud splashes as they hunt. Gray whales listen closely for the vocalizations of killer whales and swim swiftly to the safety of seaweed beds to escape these deadly enemies.

IN FOCUS

Do whales echolocate?

Baleen whales lack the oil-filled focusing melon present in other cetaceans, plus the blowhole's sound-producing "monkey lips," and the sound-conducting channel of the lower jaw that enables toothed whales to use sophisticated echolocation (the use of sound echoes to generate an "image" of the surroundings). However, minke and gray whales produce clicks that may enable them to echolocate by an unknown mechanism. The bowhead whale, which lives in Arctic waters, probably listens for echoes bouncing off the ice to navigate as it swims.

Circulatory and respiratory systems

CONNECTIONS

COMPARE the arrangement of respiratory and digestive system passages in the head of a gray whale with those of a land mammal like a ***ZEBRA*** or a ***HUMAN***.

COMPARE the diving adaptations of a gray whale with those of a ***PENGUIN*** and a ***SEAL***.

Like other mammals, a baleen whale has a four-chamber heart that pumps blood through a double circulation (the main and pulmonary circulations). Arteries with thick, muscular walls carry blood under high pressure away from the heart to supply other organs. Thin-walled veins take blood back to the heart under lower pressure. In the pulmonary circulation, waste carbon dioxide gas is released at the lungs as the red blood cells recharge with oxygen.

The heart and major blood vessels of an adult blue whale are astonishingly large. The heart weighs more than 1 ton (0.9 metric ton), with the largest artery (the aorta) and veins (the venae cavae) large enough for a person to swim inside. However, a whale's heart and vessels are no larger than would be expected for an animal of such enormous size.

Whales inhale through their nasal passages, warming the air before it travels down the trachea to the lungs. In the lungs, oxygen is exchanged for carbon dioxide, and stale air is exhaled when the whale surfaces. The respiratory system of a whale is far more efficient than that of most terrestrial mammals.

Diving

Baleen whales dive to 330 feet (100 m) or more to feed, although gray whales often make much shallower dives to the coastal seafloor. Typically, a whale breathes 10 to 15 times at the surface over three minutes or so before diving for around eight minutes. When threatened, a baleen whale can remain underwater for more than 30 minutes.

When a baleen whale dives, its lungs collapse as the surrounding pressure increases. Relative to size, a whale's lungs occupy about half the volume of those of a land mammal. Whale lungs are superefficient at extracting oxygen from air. They remove more than 80 percent of oxygen from inhaled air, as opposed to less than 20 percent in the case of humans.

IN FOCUS

Breaking the ice

Bowhead whales are right whales adapted for life in the high Arctic. They have the thickest blubber of any whale of their size, an amazingly arched upper jaw, and a powerful stocky body. The sturdy body allows them to break through sea ice up to 12 inches (30 cm) thick when they need to surface to breathe. This extends their range into frozen waters that are not accessible to other whales. Belugas benefit from the bowhead's labors and sometimes follow them to the surface. They then breathe through the newly created hole in the ice.

▼ Gray whale

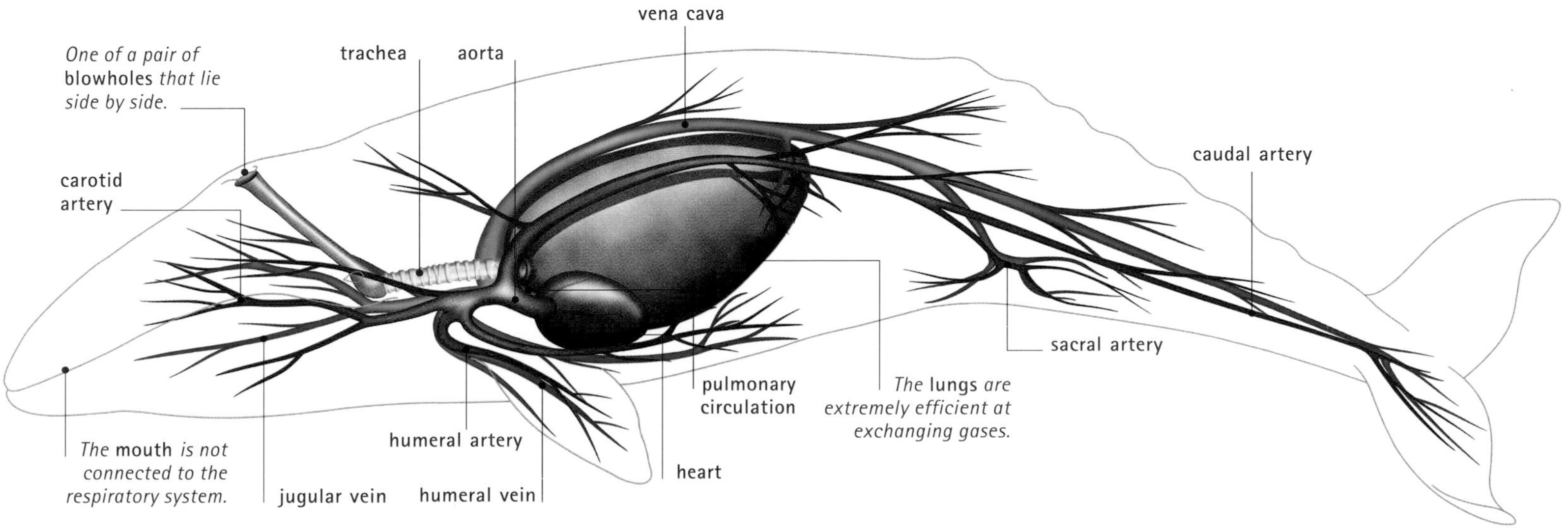

▲ *To breathe, this gray whale can breach the surface with only the top of its head. That is because cetacean nostrils have moved from the front of the skull to the top over the course of their evolution.*

Whales' blood also has many more red blood cells per unit volume of blood. Their red blood cells are larger, and overall the blood contains much more hemoglobin (the pigment that carries oxygen within the red blood cells).

When a baleen whale dives, its heart rate drops. The heart's contractions become smaller, and their frequency slows. Blood is diverted from nonessential organs to those that most need it—particularly the brain and the heart itself. Whale muscle is rich in an oxygen-trapping pigment called myoglobin. Like hemoglobin, myoglobin gradually releases its store of oxygen during the whale's dive.

CLOSE-UP

Blowholes and breathing tubes

Baleen whales have two blowholes, which are equivalent to the nostrils of land mammals. However, in cetaceans, unlike land mammals, the trachea and the esophagus (the tube leading to the stomach and intestines) are completely separate. Because of this separation, a whale can breathe only through its blowholes, not its mouth.

The bends

When a person dives deep, nitrogen gas in inhaled air gradually dissolves into the bloodstream under pressure. Then, when the diver returns to the surface, the nitrogen could bubble out of the blood. This causes a condition called decompression sickness, or the bends. The gas bubbles can block small blood vessels or gather in the gaps between bone joints to cause excruciating pain. Whales avoid decompression sickness by having compressible lungs. A whale's rib cage is flexible, and the diaphragm runs at an oblique angle. This allows the lungs to collapse readily under pressure, forcing air inside back up the trachea and into the nasal passages. These have thick, impermeable walls that prevent any gas exchange from taking place, so nitrogen cannot enter the blood when the pressure rises. When the whale moves toward the surface, the pressure lessens and the lungs reinflate.

When a baleen whale surfaces, it breathes out a foul-smelling cloud of air. This is visible above the sea surface as a spout of water, which betrays the presence of the whale. The spout is a fine spray containing seawater trapped in the blowhole and condensing water droplets from inside the whale.

Keeping warm

At 97–99°F (36–37°C), the body temperature of a baleen whale is always higher than the surrounding seawater. Whales have dense networks of blood vessels called retia mirabilia that help prevent heat wastage. Heat loss is greatest from projections that have a large surface area, such as the flippers, dorsal fin, and tail. Arteries supplying blood to these extremities are surrounded by, and run alongside, veins that carry blood back to the core. The arterial blood warms the blood traveling along the veins. This arrangement is called a countercurrent heat exchange. The extremities remain cool, while the blood returning to the core of the body stays warm.

When whales are active in warm waters there is a real danger of overheating. They avoid this by shutting down the countercurrent systems and routing blood through pathways close to the skin. This enables them to release heat into the surroundings.

The importance of the mouth

A baleen whale's tongue represents a massive surface area for heat loss. Worse, there is no insulating layer of blubber. Countercurrent heat exchange systems keep the tongue's temperature far lower than the rest of the body. In this way little heat is lost.

▼ Whales such as this gray whale need blubber and countercurrent heat exchange mechanisms to survive in cold polar waters. Migration into warmer waters necessitates other mechanisms that keep the animal from overheating.

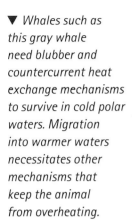

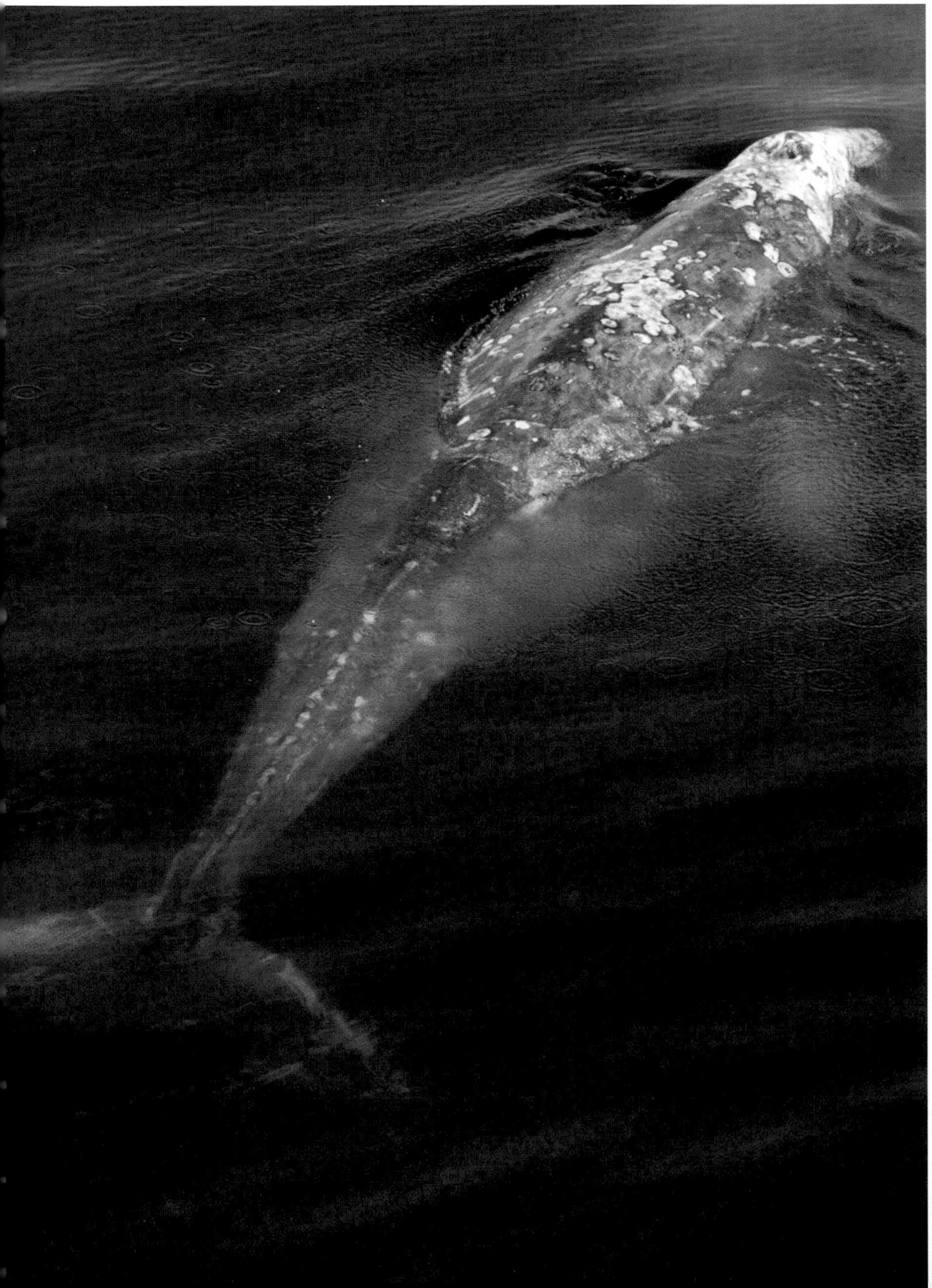

IN FOCUS

Reversing the path of heat

The countercurrent heat exchangers inside a whale's body keep the temperature constant and high. However, some parts of the body need to be cooler than others. Whales keep the temperature of these regions low by ferrying in cooler blood from other parts of the body. For example, right whales have a rete mirabile in their upper jaw. Warm blood moves in, loses its heat to the water, and then passes close to the brain. This helps keep the brain from overheating.

As in other mammals, the production and storage of sperm in whales are best at temperatures lower than the rest of the body. Whales cool their testes differently. Most adult mammals' testes lie outside the body as a result. Whale testes are inside the body; they would badly affect streamlining otherwise. In a reverse countercurrent heat exchange system, cool blood is brought from the fins to supply the testes. Female whales have similar problems. Young developing in the uterus run the risk of overheating, which would affect development. Again, cool blood from the fins passes to the uterus. It moves through the placenta, where it comes into close contact with the blood of the calf. Heat is drawn away, warming the mother's blood and cooling the calf's significantly.

Digestive and excretory systems

Baleen whales are named for their feeding structure, the baleen, which filters small organisms from the water. The size and shape of the skull and baleen plates determine the type of prey taken and the method of capture. Right whales have a very large head that approaches one-third the total length of the body. The narrow upper jaw or rostrum is curved upward and supports long baleen plates, up to 10 feet (3 m) long in northern and southern right whales, and more than 13 feet (4 m) long in bowheads. These whales feed at or near the surface, swimming along with their mouth open. The gap between the bristles is narrow; the whales sieve the water for copepod crustaceans that are less than 0.5 inch (1.25 cm) long. Occasionally, the whales feed on larger crustaceans called krill or on schools of small fish. At intervals, the whale closes its mouth and scrapes the trapped animals off its baleen with its tongue, then swallows the prey.

▶ **FEEDING**
Blue whale
Cross sections through the head of a blue whale, showing how these giant mammals use their tongue to force water through their baleen plates.

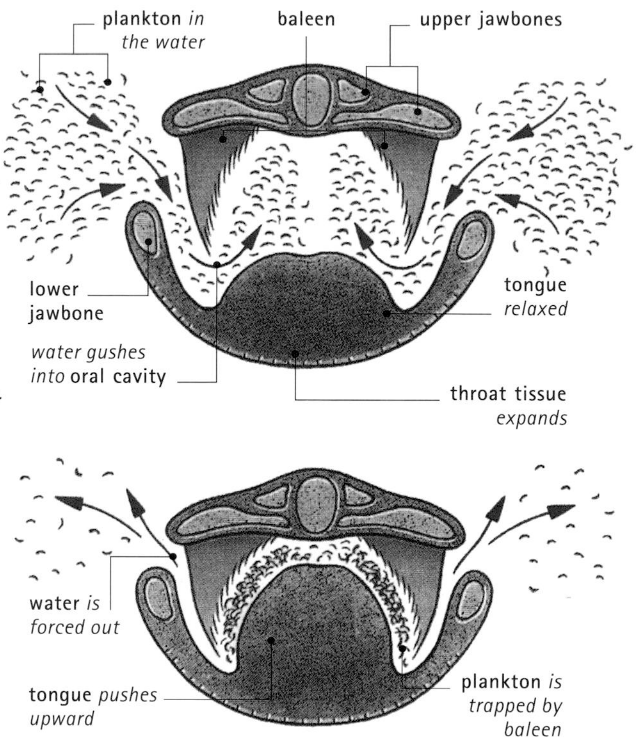
plankton in the water
baleen
upper jawbones
lower jawbone
tongue relaxed
water gushes into oral cavity
throat tissue expands
water is forced out
tongue pushes upward
plankton is trapped by baleen

1. *The whale opens its mouth, taking in a huge gulp of water containing plankton.*

2. *The mouth closes, and the tongue forces water through the baleen, straining out the plankton.*

Rorquals' feeding

Rorquals have a broader, less curved rostrum, and their baleen is shorter and bushier than that of right whales. Rorquals feed by expanding the mouth and throat cavity and engulfing a large volume of water in one go. Pleats on the underside help the throat stretch to accommodate the massive mouthful. The whale then closes its mouth and raises its tongue, forcing water through the baleen and out of the sides of the mouth. Small prey items are trapped on the inside of the baleen plates, scraped off with the tongue, and swallowed.

Rorquals' main prey items vary from species to species, depending on feeding method, locality, and the fineness of the bristles on the

▼ **Gray whale**

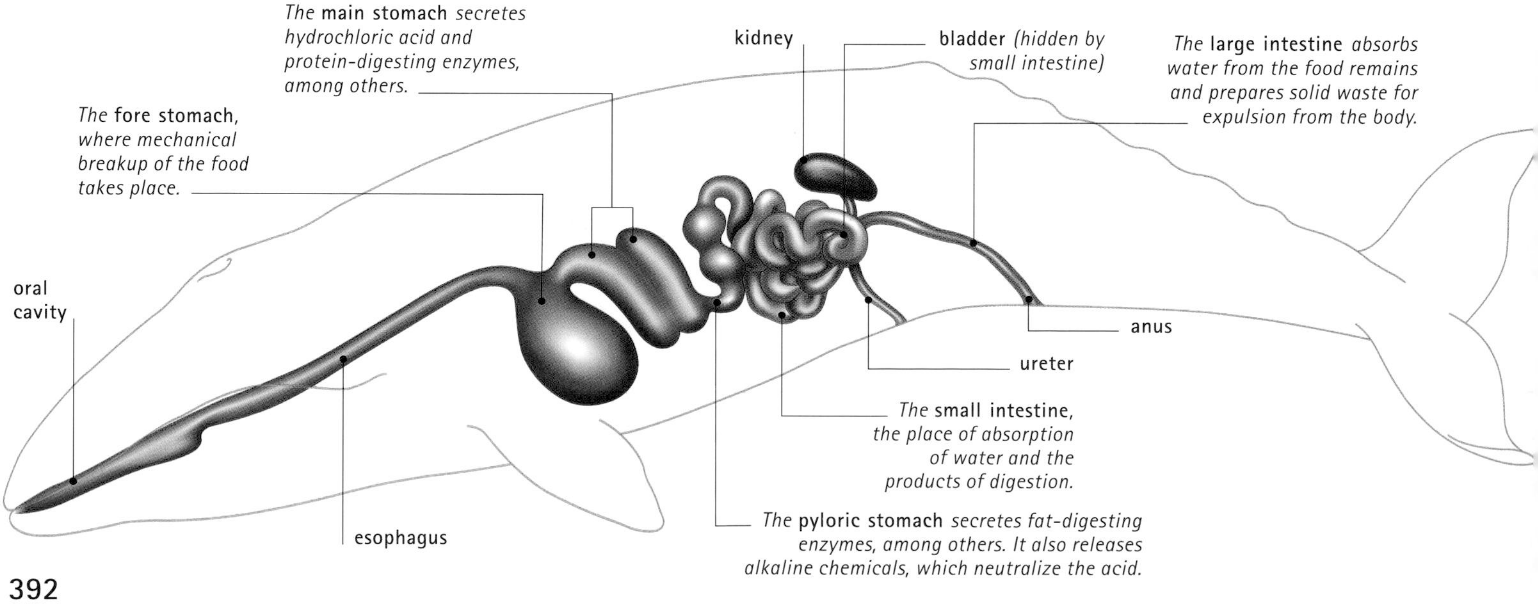
The main stomach secretes hydrochloric acid and protein-digesting enzymes, among others.
kidney
bladder (hidden by small intestine)
The large intestine absorbs water from the food remains and prepares solid waste for expulsion from the body.
The fore stomach, where mechanical breakup of the food takes place.
oral cavity
anus
ureter
The small intestine, the place of absorption of water and the products of digestion.
The pyloric stomach secretes fat-digesting enzymes, among others. It also releases alkaline chemicals, which neutralize the acid.
esophagus

baleen plates. Sei whales have fine-bristled baleen that captures planktonic crustaceans, especially copepods. The blue whales' baleen is medium-bristled; its preferred diet is krill. Fin and humpback whales have coarser baleen plates, and they often feed on shoals of herring and capelin fish.

Plowing the seabed

Uniquely among baleen whales, gray whales feed at the seabed. They have stout jaws with short, stiff baleen up to 10 inches (25 cm) long. In their northern feeding grounds, gray whales plow the seabed with their baleen, sieving out amphipod crustaceans and a variety of other invertebrates including clams and worms. They also sieve the water for disturbed invertebrates after most of the sediment has settled. The gray whale's tongue is unusually large. It aids in scooping up sediment and manipulating the mouthful to help separate food from mud.

On migration, gray whales occasionally feed in a manner similar to other baleen whales. They capture small fish at near the surface and take shrimplike mysid crustaceans that thrive in beds of kelp (large brown seaweeds).

Digestion and absorption

Swallowed food travels down the esophagus to the stomach, where digestion begins. Whale stomachs have three compartments. The first is the fore stomach, where food is ground up into a soup. This is squirted into the second stomach chamber, the main stomach. The walls of this chamber secrete hydrochloric acid, and protein-digesting enzymes that chemically break down the food. The walls of the third stomach compartment, the pyloric stomach, secrete fat-digesting enzymes, more protein-digesting enzymes, and an alkaline fluid that neutralizes the acidity of the juice from the main stomach. From the stomach, the food enters the first part of the small intestine. The intestine walls absorb the products of digestion, which enter the bloodstream before circulating to the tissues that need them. The small intestine walls are highly folded and rich in blood vessels. This provides a very large surface area across which digested food and water can be absorbed.

CLOSE-UP

Big appetites

The fore stomach of a blue whale can contain about 1.1 tons (1 metric ton) of krill. At the height of the summer feeding season, the whale consumes about 4.4 tons (4 metric tons) of food each day. That is equivalent to the weight of a small truck.

▲ *This gray whale is feeding on krill at the ocean surface, a behavior these whales exhibit during their migration from feeding to breeding grounds. Note the wispy edges of the baleen plates that are visible in the animal's mouth.*

Gaining water and losing salt

The salt concentration of baleen whale blood is considerably lower than that of seawater. Whales must take in water to keep their blood diluted; they gain much of their water from food. Fish have a salt concentration similar to a whale's own tissues, but invertebrates, such as krill and copepods, have tissues with a salt concentration similar to that of seawater. To get the water they need, whales take on unwanted salts that they must excrete.

Whale kidneys remove excess salts from the bloodstream, along with other wastes. Water, salts, and the other wastes collect in the urine. This travels from kidneys to bladder before being expelled through the urethra. Perhaps surprisingly, whale kidneys are not very efficient. They produce urine that is more concentrated than seawater, but they waste large amounts of water that must be replaced by feeding and drinking.

CONNECTIONS

COMPARE the feeding strategy of a gray whale with that of a toothed whale such as a ***DOLPHIN***.

COMPARE the three-chamber stomach of a whale with the four-chamber stomach of a ruminant such as a ***GIRAFFE***.

Reproductive system

In many ways, the baleen whale reproductive system is very similar to that of any other placental mammal. However, there are some major differences caused by the need for streamlining, and because mating, giving birth, and suckling must take place underwater.

Male cetaceans have an internal penis and testes to aid streamlining. The penis normally lies inside the abdomen. Before mating, it fills with blood then emerges through the genital slit. Unusually, a muscle attached to the penis makes it relatively mobile, and retracts it back into the abdomen after mating.

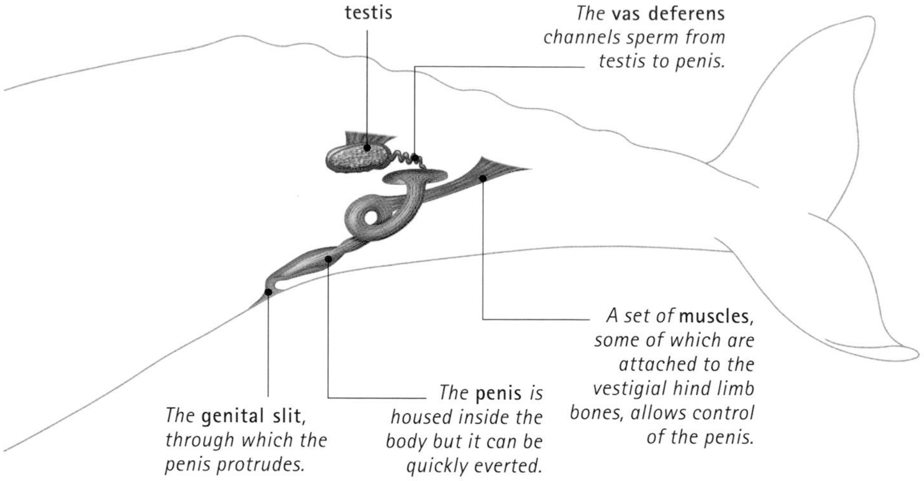

testis
The vas deferens channels sperm from testis to penis.
A set of muscles, some of which are attached to the vestigial hind limb bones, allows control of the penis.
The penis is housed inside the body but it can be quickly everted.
The genital slit, through which the penis protrudes.

▲ **MALE REPRODUCTIVE SYSTEM**
Gray whale
The flexible penis allows great maneuverability during copulation and in other social interactions.

Mating whales

In humpback whales, more than one male will usually compete to mate with a single female. Males advertise their presence by singing mating songs that also serve as challenges to other males. Male and female gray whales caress each other in a courtship ritual before mating (copulation) begins. The male's penis is flexible and about 6 feet (2 m) long. It releases sperm into the female's vagina; copulation may last only 20 to 30 seconds, but it may be repeated on several occasions.

Most baleen whales are probably polygamous—males and females each have several partners. Often, males compete with one another over a female, using their physical bulk to get between the receptive female and rival males. Except in species where several males coerce a female, such as right whales and the gray whale, female whales are not submissive but actively select a mate.

Pregnancy and birth

In baleen whales, the gestation period (the time from copulation to birth) is between 10 and 13 months, which is no longer than that of the much smaller toothed whales. Adult female gray whales breed only once every two or three years and typically bear one calf at a time. Early whalers nicknamed gray whales "devilfish," after experiencing the wrath of mother whales. They would attack boats when they or their calves were threatened.

Rarely observed, the birth of a baleen whale usually takes place at or near the surface, with the mother helping the calf to the surface to take its first breath. Baleen whale calves are usually born headfirst. However, baleen whales, including gray whales, have been observed giving birth tailfirst, as is generally the case in toothed whales.

Becoming independent

All traces of the mother's milk-producing mammary glands are usually hidden, but when nuzzled by a calf, a nipple emerges. Whale calves lack the mouth shape and musculature to be able to suck. Instead, they squeeze the mother's nipple between their tongue and the roof of their mouth. In response, muscles

▼ *Mating gray whales. The long, pink organ is a male's penis. Several males attend to a female when she is in estrus.*

surrounding the mammary glands contract and squeeze, squirting milk directly into the calf's mouth. A calf stays close to its mother during the first few weeks of its life and suckles regularly. It puts on weight at an incredible rate—faster than the young of any other mammal. A blue whale calf is about 23 feet (7 m) long at birth and weighs about 2 tons (1.8 metric tons). It grows at a rate of about 180 pounds (85 kg) each day, reaching 25 tons (23 metric tons) and 48 feet (15 m) long by the age of eight months. Whale milk is 40 percent fat or more. By comparison, cow's milk is between 3 and 5 percent fat. A high fat intake enables a whale calf to develop a thick layer of blubber within weeks.

The calves of rorqual whales begin to eat solid food within six or seven months. All baleen whale calves are fully weaned within 12 months. Gray whales become sexually mature at about eight years old, when males are about 36 feet (11 m) long and females about 38 feet (11.5 m). Gray whales can live to 40 years or more, and some bowhead whales can reach an astonishing 150 years old.

Migration

Gray whales and almost all the larger rorquals make long-distance migrations between summer feeding grounds in polar or subpolar waters and their winter breeding grounds in tropical or warm temperate seas. During these migrations, which in the case of the gray whale involve a round trip of more than 11,000 miles (18,000 km), the whales use a great deal of energy. Why do they undertake these long journeys? Earth's most food-rich waters occur in polar waters in summer, where the combination of continuous sunlight and abundant nutrients encourages an explosion of plantlike phytoplankton. They support vast numbers of zooplankton, on which fish and many other animals feed. Baleen whales feed in polar waters to reap this rich harvest, allowing them to maintain their massive body weight. By migrating during the leaner winter, adults ensure that their calves are born in warmer waters. Calves begin life with only a thin insulating layer of blubber. They can build up a thicker layer by the time they have completed their journey to chilly polar waters.

TREVOR DAY

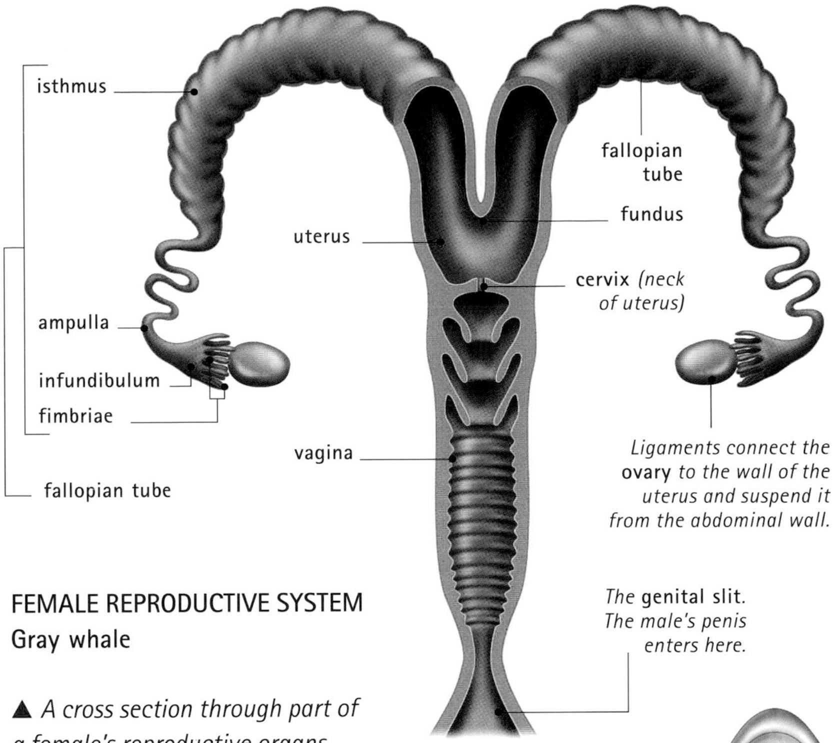

FEMALE REPRODUCTIVE SYSTEM
Gray whale

▲ *A cross section through part of a female's reproductive organs.*

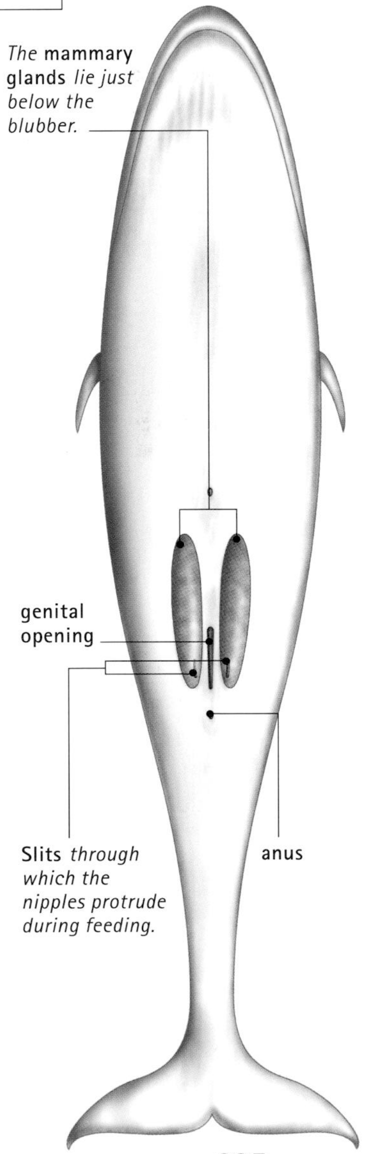

The mammary glands lie just below the blubber.
genital opening
Slits through which the nipples protrude during feeding.
anus

▶ *The mammary glands of a female gray whale are astonishingly productive. They can deliver up to 50 gallons (190 liters) of extremely rich milk each day.*

IN FOCUS

Competing sperm

Male northern and southern right whales have the largest sperm-producing organs, or testes, in the animal kingdom. Each pair weighs about 1 ton (0.9 metric ton). They release tens of gallons of sperm during mating. In these species, and in gray whales, which have similarly massive pairs of testes, males do not compete for the attentions of a female. Instead, they actively help each other out. Several males mate with any one female. Rather than the whales actively competing, their sperm do battle inside the female instead, in the race to fertilize her ovum (egg). This is called sperm competition.

FURTHER READING AND RESEARCH

Mead, James G. and Joy P. Gold. 2002. *Whales and Dolphins in Question: The Smithsonian Answer Book.* Smithsonian Books: Washington, DC.

Green anaconda

ORDER: Squamata SUBORDER: Serpentes
FAMILY: Boidae SPECIES: *Eunectes murinus*

There are 69 species of boas and pythons. Most live in the tropics and subtropics, although a few live in temperate climates. They dwell on all continents except Antarctica. Some are very large; the largest, the green anaconda, is the heaviest snake in the world. Others, such as the rubber boas, are much smaller.

Anatomy and taxonomy

The relationship between organisms is largely determined by studying features of their anatomy, although genetic research is also important. Boas, pythons, and all other snakes are reptiles. Other reptiles include lizards, crocodiles, turtles, and tuataras. Many other types of reptiles lived millions of years ago, including the dinosaurs.

- **Animals** All animals are multicellular (many-celled) and depend on other organisms for food. They differ from other multicellular life-forms in their ability to move around and in their rapid response to stimuli.

- **Chordates** At some stage in its life a chordate has a stiff, dorsal (back) supporting rod called a notochord.

- **Vertebrates** In vertebrates, the notochord develops into a backbone or spine made up of units called vertebrae. Vertebrates have a system of paired muscles lying on either side of a line of symmetry.

- **Reptiles** Reptiles are four-legged vertebrates (although some have lost their legs during the course of evolution) that have a thick, horny skin, usually divided into plates called scales. Most reptiles lay eggs, but in some species the eggs are retained within the body of the female until they are ready to hatch. Most reptiles are unable to generate their own body heat as birds and mammals do. They can, however, regulate their body temperature by going into hot places when they need to warm up and cooler places when they need to lose heat.

- **Squamates** Lizards and snakes are classified together to form the order Squamata. Squamate skulls are not as solid as those of turtles and crocodilians. Male squamates also have two penises.

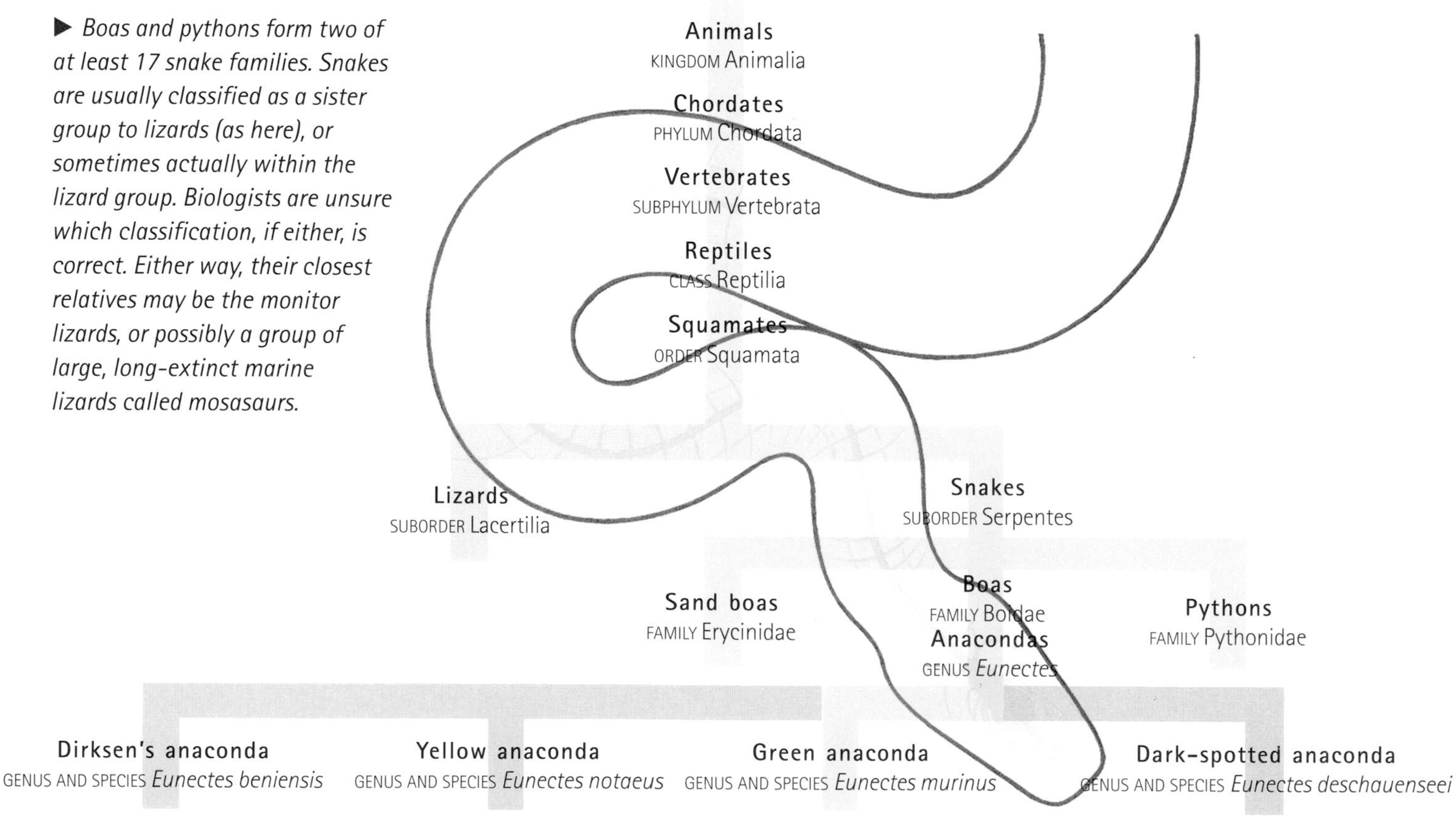

▶ *Boas and pythons form two of at least 17 snake families. Snakes are usually classified as a sister group to lizards (as here), or sometimes actually within the lizard group. Biologists are unsure which classification, if either, is correct. Either way, their closest relatives may be the monitor lizards, or possibly a group of large, long-extinct marine lizards called mosasaurs.*

▲ *A green anaconda subdues its prey, a caiman. The snake tightens its grip each time the caiman breathes out.*

• **Snakes** Snakes do not have legs, but neither do many kinds of lizards. Snakes can be distinguished from lizards because they usually have only a single row of scales (the ventral scales) on their underside. Their eyes are covered by a transparent scale, and they cannot close their eyes or blink.

• **Boas** There are 29 species in the boa family, including the boa constrictor and the green anaconda. Like their close relatives the pythons, boas usually kill their prey by winding coils of their body around it and squeezing hard, eventually causing death by suffocation. This is called constriction. Neither boas nor pythons are venomous. Boas live in Central and South America, some islands in the Caribbean, Madagascar, and New Guinea.

• **Anacondas** There are four anaconda species. One was named by biologists as recently as 2002. Only the green anaconda grows to a huge size. It lives in South America, from Venezuela and Colombia south to Argentina.

FEATURED SYSTEMS

EXTERNAL ANATOMY Green anacondas are long but stout compared with other snakes. Their body is covered with scales. *See pages 398–401.*

SKELETAL SYSTEM Anacondas have more than 300 vertebrae in their backbone; all except the first and those at the tip of the tail have ribs attached to them. Many of the jawbones are only loosely connected to one another, allowing individual bones to move apart when the animal is swallowing large prey. *See pages 402–404.*

MUSCULAR SYSTEM Powerful muscles running along the body enable anacondas to move on land and swim in water by bending the body laterally. They also provide the powerful squeeze used to constrict prey. *See page 405.*

NERVOUS SYSTEM The eyes of snakes focus in a different way from those of other vertebrates. Anacondas can sense heat radiation using sense organs in the jaws, helping them detect prey. *See page 406.*

CIRCULATORY AND RESPIRATORY SYSTEMS Anacondas, like all snakes, have a three-chamber heart and two main systemic arteries. The left lung is tiny and does not function as a respiratory organ. *See page 407.*

DIGESTIVE AND EXCRETORY SYSTEMS The stomach is large, enabling anacondas to digest very large prey. Prey is swallowed whole. *See page 408.*

REPRODUCTIVE SYSTEM An adult female anaconda retains her eggs within the oviducts. The eggs hatch inside the female, which effectively gives birth to live young. Male anacondas have two penises. *See page 409.*

External anatomy

COMPARE the camouflage skin pattern of a green anaconda with the color-changing camouflage of a ***JACKSON'S CHAMELEON***.

COMPARE the position of a green anaconda's eyes with the position in a ***CROCODILE***. Both these animals spend a lot of time in water with their head at the surface.

Green anacondas are the heaviest snakes in the world. The biggest weigh over 440 pounds (220 kg). These are always females—male anacondas are much lighter. Adult females may grow to a length of over 30 feet (10 m). Many exaggerated claims have been made about how long anacondas can grow. If the skin of a dead snake is removed, it expands as it dries, so after a while the total length of the skin is greater than it was in life. This may lead to inaccurate estimates of length. In comparison with many other snakes, anacondas have a stout appearance. The snout of an anaconda is relatively pointed and the head has a smaller diameter than the body. This helps to streamline the animals while they are swimming. Anacondas spend a lot of time in water, but they also sometimes climb trees.

Green anacondas live for between 10 and 30 years in captivity. The young grow rapidly and can begin breeding when they are three to four years old. If an individual survives to a large size in the wild it will probably live for 10 to 15 years.

Camouflage **skin** *patterns hide the anaconda from both predators and prey.*
The pupils narrow to slits during the daytime to protect the **eyes** *from the sun but they widen at night, allowing as much light as possible to enter. The eyes are positioned on the top of the head, allowing the snake to see when lying at the surface of water.*
Like the eyes, the **nostrils** *are also positioned near the top of the head, enabling the snake to breathe while largely submerged in water.*
The **stripe** *running from the rear of the head to the eye helps to disguise the eye's circular shape, making it more difficult for prey to detect the snake as it lies in wait to attack.*

▶ *Characterized by a camouflage skin pattern and a long, broad, and powerful body, the green anaconda is a formidable predator.*

Sit and wait

An anaconda's ears, like those of all snakes, are internal and cannot be seen from the outside. Their eyes and nostrils are nearer the top of the head than in many snakes. This allows them to breathe while lying with their head at the surface of the water looking for prey.

Anacondas are "sit-and-wait" predators. They do not spend much time actively searching for food but lie quietly in wait and then pounce on prey when it passes. Their green-and-brown skin pattern camouflages them from prey.

Anacondas feed mostly at night. During the day the pupils of their eyes are vertical slits but at night they open up, allowing as much light as possible to enter. Adult green anacondas

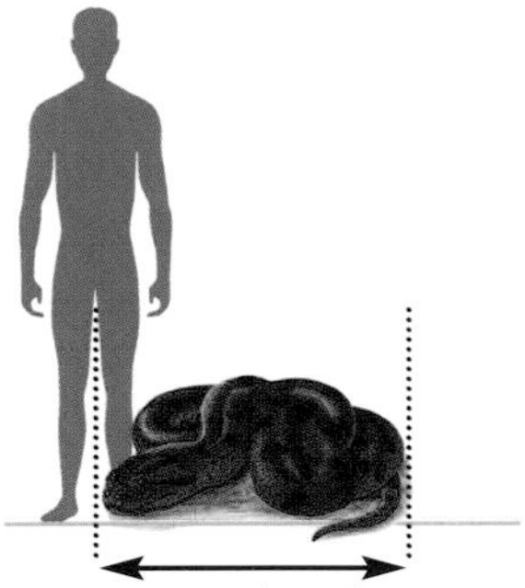
30 feet (10 m)
when extended

Scales *cover the body of the green anaconda, providing protection from both predators and the environment.*

COMPARATIVE ANATOMY

Snake eyes

The eyes of snakes are covered with a transparent scale called the spectacle scale. Snakes do not have movable eyelids and cannot blink or close their eyes. In many snakes the pupils of the eyes are round both during the day and at night. Other snakes, including the green anaconda, have pupils that look like thin vertical slits by day. These snakes hunt at night or vary the time of day when they are looking for food. Many snakes in temperate climates, for example, hunt at night during the summer. During spring and fall, however, it is too cold at night for them to be active, so they hunt during the daytime.

The pupils of night-hunting snakes open wide at this time to let as much light as possible into the eyes. The image is focused by the lens onto an area at the back of the eye called the retina, which is very sensitive. The retina would be damaged if too much light fell on it, so the pupils narrow down to slits during the day. This narrowing limits the amount of light that gets through.

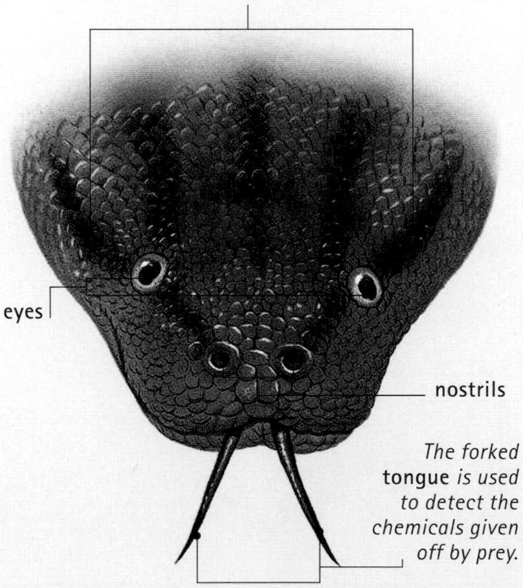
camouflage stripes
eyes
nostrils
The forked **tongue** *is used to detect the chemicals given off by prey.*

Arafura file snake

In contrast to snakes such as anacondas, which have eyes that narrow to slits during the daytime, the Arafura file snake has eyes that are round both during the day and at night.

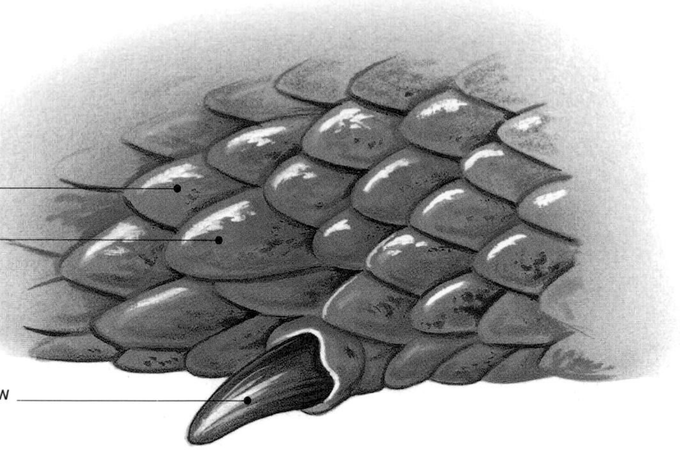

▲ **PELVIC CLAW**
The pelvic claws found on boas and pythons appear as small spurs on the rear underside of the snakes. The spurs probably no longer serve a function.

feed on deer, capybaras (large rodents that are common in parts of South America), caimans (kinds of crocodiles), peccaries (a group of piglike mammals), and other large animals. Smaller anacondas eat a wider range of prey, and fish make up much of their diet. Anacondas rarely attack people in the wild, but captive anacondas are often aggressive. They bite and may attempt to constrict their keepers.

The teeth are pointed and are angled backward on the jaws. Unlike many snakes, anacondas do not have fangs. An anaconda's tail is short and much thinner than the body. Just in front of the tail, on either side of the body, are two little claws. The ancestors of anacondas had hind limbs, of which these claws are all that remains. Boas and pythons are the only snakes that still have these claws.

COMPARATIVE ANATOMY

Colors of snakes

Some snakes are very brightly colored. If they are venomous, the bright colors may warn a predator to leave them alone; this is called warning coloration. Some nonvenomous snakes also bear bright warning coloration; they are called mimics. Many snakes have bars of dark color running from side to side or lines of dark or pale color running along the length of their body. These may confuse predators, making it difficult for them to figure out how fast the snake is moving. Snakes such as anacondas avoid predation with camouflage skin patterns. One of the most conspicuous parts of the body in many animals is the eye. Anacondas, like many camouflaged animals, have a dark stripe leading up to each eye. This helps disguise the eye's circular shape.

CLOSE-UP

Big females, smaller males

▲ *The size difference between males and females is particularly apparent during mating, when many small males try to mate with one female.*

Differences in the appearance of adult males and females in a species are called sexual dimorphism. Male lizards and birds, for example, are often more brightly colored than females. Most snakes do not have much sexual dimorphism in color or pattern, but the majority do have a discrepancy in size. Females are almost always bigger than males.

Females are bigger because they need a lot of space for oviducts. Eggs are stored in these tubes while the clutch is developing. Some snakes are ovoviviparous; the eggs develop and hatch inside the female. Females of species that give birth to live young need to store the clutch inside them for long periods. In the case of female green anacondas, this storage can last for six months.

The size difference between female and male green anacondas is greater than in any other snake species. While there has been a huge amount of attention given to the size that can be achieved by females, the more modest size of male anacondas has not been studied in detail.

CLOSE-UP

Skin form and function

The outer layer of a snake's skin is made up of thick horny plates called scales. The skin between the scales, which is softer and not so thick, is often hidden by the overlapping scales. The outer layer of the skin is made up of dead cells. Underneath this layer are living cells. In order to grow, a snake must replace its outer layer of skin. At intervals, therefore, the outer layer of all snakes peels off. This process of shedding the outer skin is called sloughing. Prior to sloughing, the skin color of a snake becomes dulled. Snakes usually shed the entire skin in one piece. The old skin splits at the front of the head, and the snake then glides out. Shed skins can sometimes be found lying around where snakes have been living. The scales are so distinct that it is sometimes possible to determine what species of snake the skin came from.

► *The overlapping scales of the green anaconda. Contrary to popular belief, snakes do not feel slimy. The scales make the snake dry to the touch.*

Scaly skins

The body surface of anacondas is covered by plates of thickened skin called scales, which provide a protective surface. On the back, sides, and underside of the body, each scale slightly overlaps the scale behind it, but most of the scales on the head and the tail do not overlap in this way. The scales on the underside of the anaconda's body, where the snake is in contact with the ground, are larger than any others. They are called ventral scales, and there are between 246 and 259 of them. Each ventral scale corresponds to a single vertebra of the spine. The ventral scales of anacondas and their relatives are not as wide as those of most other snakes. Along the underside of the tail are between 67 and 71 subcaudal scales. In boas, including green anacondas, the subcaudal scales form a single row. This is unusual. In the closely related pythons and most other snakes, the subcaudals are paired. The difference in subcaudal scales is the most important external feature distinguishing boas from pythons.

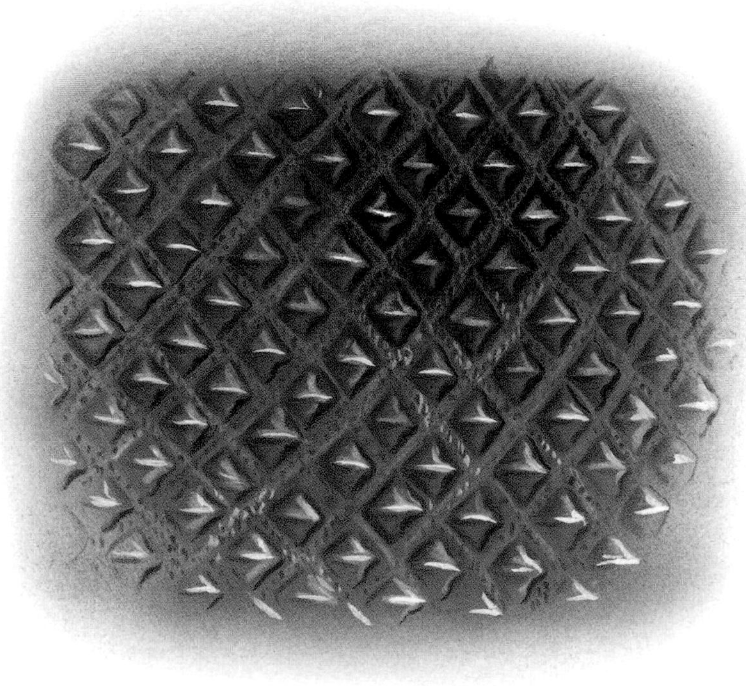

▼ SCALES
File snake
The ventral (underside) scales of an aquatic snake, such as an Arafura file snake, are relatively small. Aquatic snakes do not need large ventral scales to assist locomotion, since they rely on swimming.

Scale differences

The scales along the anaconda's lips are also large. Those along the upper lip are called supralabials, and there are 15 to 17 of them on each side. Those along the lower lip are called infralabials and number 20 to 22 on each side. The scales on the head behind the eye are about the same size as the scales on the top of the body. In many snakes the scales on this part of the head are much larger and are called shields. Like boas, however, vipers and rattlesnakes lack the larger shields at the back of the head. The scales in front of the eyes are larger than those behind.

Skeletal system

Because snakes are long and thin and have no limbs, their skeletal systems are very different from those of their four-limbed relatives. The spine has a very large number of vertebrae (backbones). Anacondas have between 313 and 330 vertebrae, and some snakes have more than 400 (by comparison, humans have just 32 vertebrae). The points where one vertebra is connected to the next by cartilage are called articulations. Most reptiles have three of these at the front and the back of each vertebra. If snakes had only three articulations, the long thin backbone would be loose and would be prone to twisting. Snakes, however, have five articulations at the front and back of vertebrae. The main one is the condyle, a knoblike projection at the back of each vertebra that fits into a cup on the front of the vertebra behind. Above the condyle and cup are two thin projections called zygapophyses. Each backward-projecting zygapophysis at either side of a vertebra articulates with the forward-projecting zygapophysis from the vertebra behind it. All reptiles have condyles and zygapophyses. Snakes, in addition, have two forward-facing projections, one on each side near the top of each vertebra. These articulate with small sockets at the back of the vertebra directly in front of them. They are called zygosphenes and zygantra. All snakes have zygosphenes and zygantra. A few lizards also have them, although most do not.

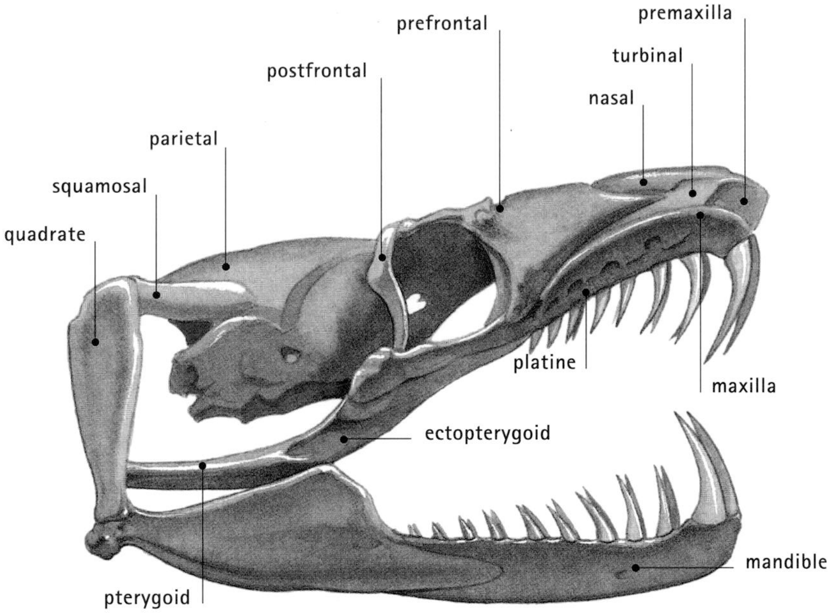

▲ **SKULL**
Green anaconda
An anaconda's upper and lower jawbones are only loosely connected, enabling them to open very widely to swallow large prey.

▼ *A green anaconda's vertebrae have more connecting points, or articulations, than those of most other animals. These extra articulations help prevent the long thin backbone from excessive twisting.*

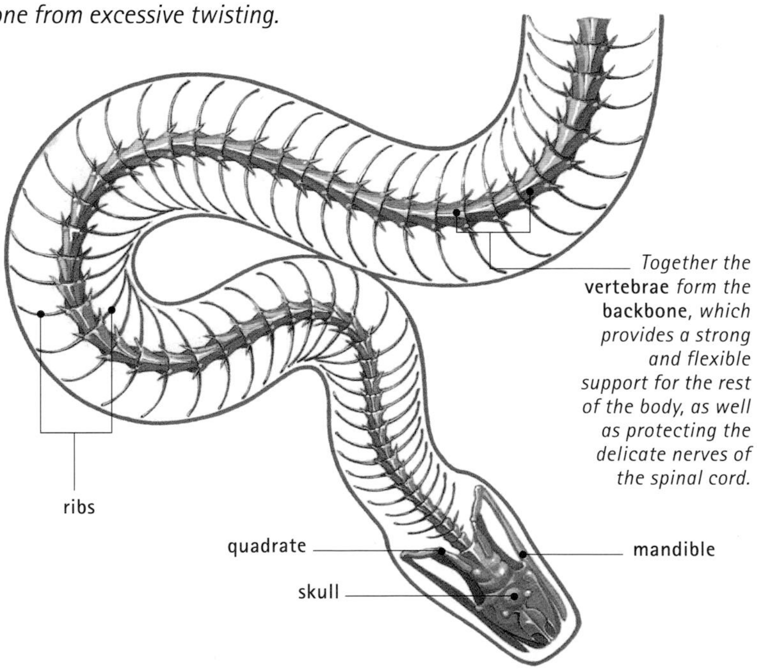

Together the vertebrae form the backbone, which provides a strong and flexible support for the rest of the body, as well as protecting the delicate nerves of the spinal cord.
ribs
quadrate
mandible
skull

Snake ribs

An unusual feature of snake skeletons is that most of the vertebrae have ribs. The first vertebra does not, nor do the vertebrae at the far end of the tail. These do, however, have projections that resemble small ribs. The attachments of the ribs are very loose. At their top ends, where they articulate with the vertebrae, the ribs can move in forward and backward directions. At their far ends they are attached by cartilage to the inner parts of their corresponding ventral scales.

Vestiges of ancient limbs

Since they have no legs, snakes have no need for shoulder or hip girdles. Green anacondas are one of the few kinds of snakes that still have hip girdle bones. On each side of the body the three bones that make up the hip girdle can be recognized: the pubis, ilium, and ischium. The pubis bone is long and thin, and points toward

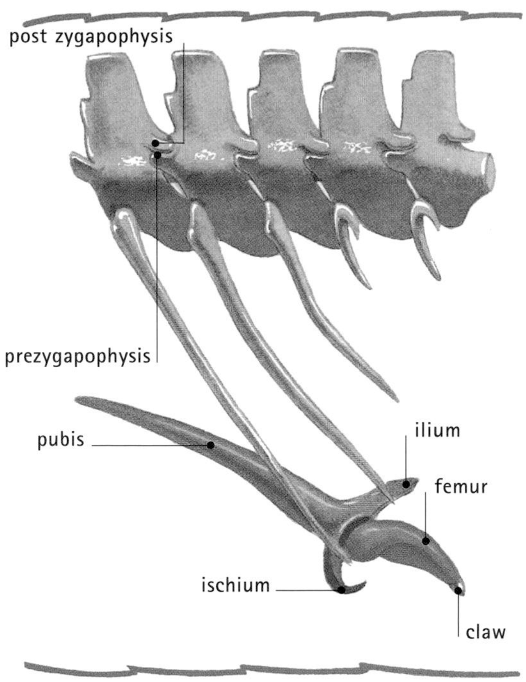

▲ Vestigial hind leg and pelvis
Anatomical structures that no longer have any practical function are known as vestigial. Boas and pythons have vestigial hind legs and a pelvis, which indicate that their ancestors had limbs.

the head. The ilium, which points backward, is much shorter, and the ischium is the smallest of the three bones. In vertebrates with limbs, the top bone of the hind leg (called the femur) articulates with the hip girdle through a ball-and-socket joint. In green anacondas the femur is just a small piece of cartilage. At its farthest tip this has a small horny claw. The two claws (one on each side of the body) project through the skin, emerging on either side and just in front of the opening for the sexual organs (the vent). The hip girdle is not attached to the backbone, as it is in vertebrates with limbs.

Mobile skull bones

The loose arrangement of the bones in the skull of snakes enables them to open their jaws very wide so they can swallow extremely large prey. Snakes do not break their prey into pieces; they swallow it whole. The most important teeth on each side of the upper jaw are those fixed to the maxilla. There are also teeth attached to the palatine and pterygoid bones, forming an additional row on each side at the top of the mouth. There are four rows of

COMPARATIVE ANATOMY

The skulls of burrowing snakes

Although the skull of all snakes is built on the same basic plan, there is some variation in the details. Blind snakes and thread snakes spend their lives burrowing in the soil, searching for the earthworms, termites, and burrowing lizards that make up most of their prey. They have a solidly constructed braincase, which enables the head to be used as a "battering ram" that can be forced through soil. Their jawbones are not as mobile as those of other snakes. Thread snakes have teeth only on the lower jaw; blind snakes have teeth only in the upper jaw. Like blind snakes, pipe snakes spend a lot of time burrowing; they have a solid skull, firmly connected jawbones, and fewer teeth than most other snakes.

Blind snake skull

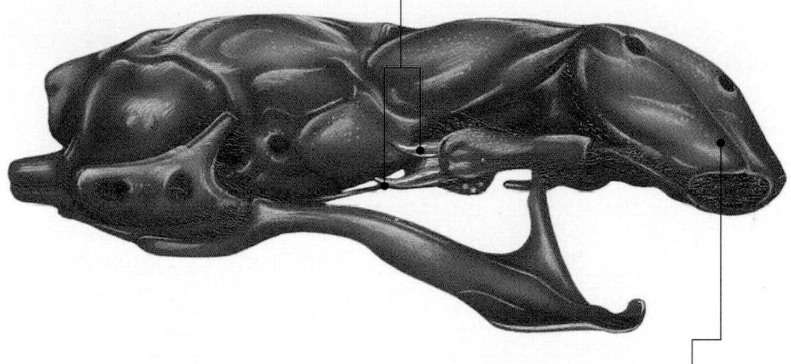

Thread snake skull

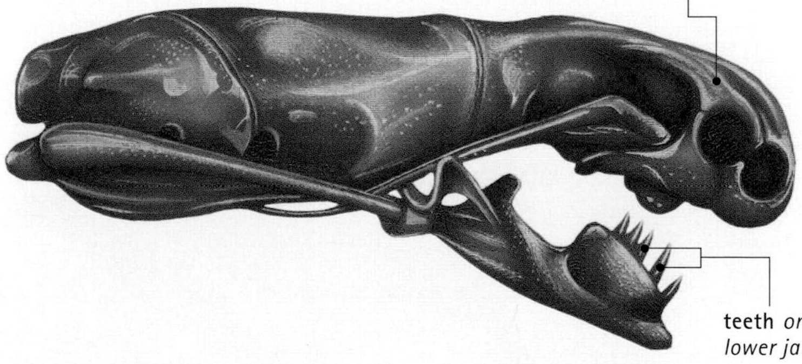

Sri Lanka pipe snake skull

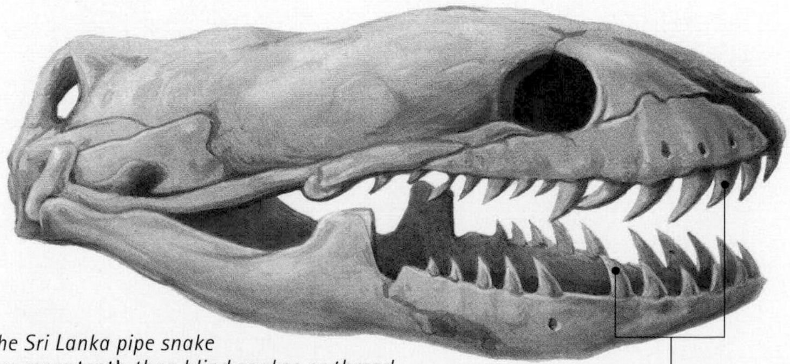

The Sri Lanka pipe snake has more teeth *than blind snakes or thread snakes but fewer than most other snakes.*

CONNECTIONS

COMPARE the loosely connected jawbones of a snake with the rigidly connected jawbones of an insect-eating animal, such as a ***GIANT ANTEATER***, which does not need to open its mouth wide to swallow its prey.

teeth at the top of the mouth. There are four rows of teeth in total in the upper jaw. The teeth of the lower jaw are attached to the mandible bone; there are only two rows of these teeth.

Many of the bones in the jaw are able to move relative to one another. The jaw has powerful muscles that make this possible. When a green anaconda opens its jaws, the long, thin pterygoid and ectopterygoid bones of the upper jaw are pulled forward. This action pushes the maxilla and mandible bones forward. The connective tissue joining all of these bones can stretch, allowing the bones to move independently of each other. The two maxillae are joined by connective tissue in the snout, so these bones can also be pulled apart.

As the snake closes its mouth, the pterygoid, mandible, and maxillae are pulled backward. This movement causes the teeth to move backward, pulling prey farther into the mouth and toward the gullet. Similar movements of other jawbones help this process. The movements of bones on either side of the mouth and in the upper and lower jaws can take place independently of each other, so a snake can pull the teeth backward on one side of the jaw while pushing the teeth forward on the other.

COMPARATIVE ANATOMY

Fang action

In many snakes, the salivary glands in the upper jaw produce venom. Cobras, mambas, and their relatives have a pair of enlarged, curved teeth at the front of the maxilla. These are the fangs. There may sometimes be several smaller pairs of fangs, but those are "spares" that will grow to their full size only if the main pair gets damaged. Vipers and rattlesnakes have fangs that can move. When the jaws are closed the fangs lie facing backward along the upper jaw. When the mouth is opened, muscles in the upper jaw pull the pterygoid and ectopterygoid bones forward. The maxillae, bearing the fangs, are very short. Forward movement of the ectopterygoid bones causes the maxillae to rotate; the fangs move around with them so that they are pointing forward. Viper and rattlesnake fangs are hollow. The venom is injected into the prey through a hole at the tip of each fang. When the snake starts swallowing its prey, the fangs lie flat along the jaws. Anacondas do not have fangs, so they are not venomous.

► Egyptian cobra

Cobras have small fangs at the front of the upper jaw. Like many other snakes, cobras are able to swallow large prey owing to the elasticity of the connections between the upper and lower jawbones.

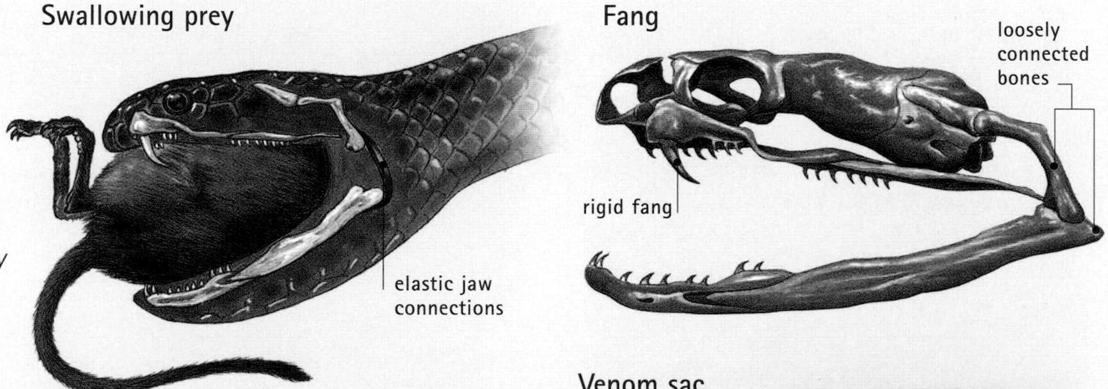

► Bibron's burrowing asp

Burrowing asps are a kind of viper and so have hollow fangs through which venom is injected. The snake stores the venom in a venom sac in the rear of the head. When not in use the fangs lie flat inside the mouth.

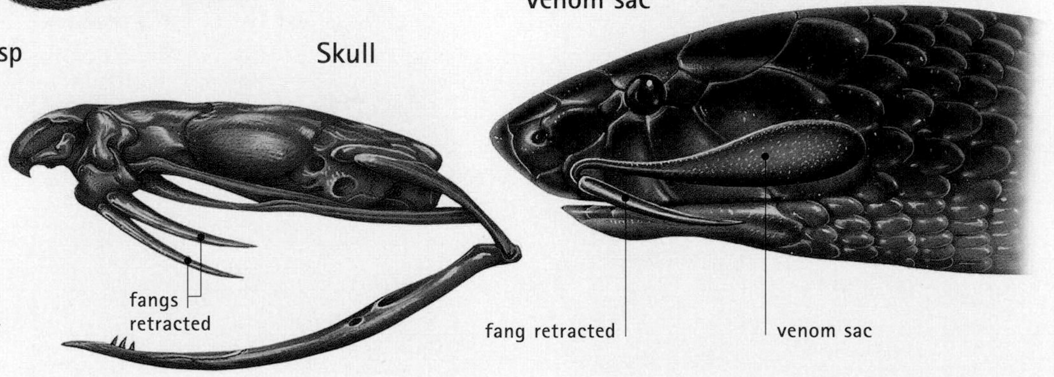

Muscular system

The arrangement of muscles in a snake is very different from that in other vertebrates. Movement, both on land and in the water, is achieved by waves of contraction that pass backward along one side of the body and then the other. These bend the body so that undulations pass backward, and the force exerted by these on the water or earth causes the snake to move forward. The muscles that run along the backbone and along and around the body, and those that attach to the ribs, are all extremely complex.

The biggest muscles run on either side of the backbone. The three most powerful muscles running along the body are the longissimus dorsi, the spinalis, and the iliocostalis. These muscles are very important for snakes like anacondas, which constrict their prey before swallowing it.

Jaw muscles

The jaws also have powerful muscles and they, too, are very complicated. The most important are the temporalis anterior muscles, which provide most of the power for closing the jaws. In vipers and rattlesnakes the protractor-pterygoideus muscles are well developed: they move the pterygoid bones forward, causing the fangs to be erected.

▼ HEAD MUSCLES

The jaw muscles of a green anaconda must have the strength to bite and hold onto prey, while also possessing the flexibility to allow the jaws to open very wide for swallowing.

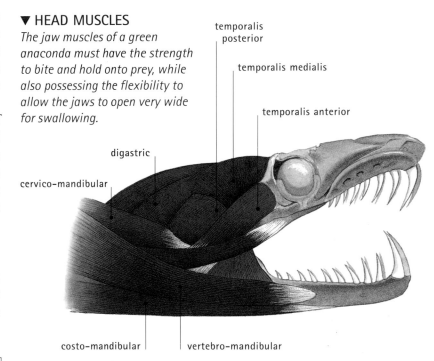

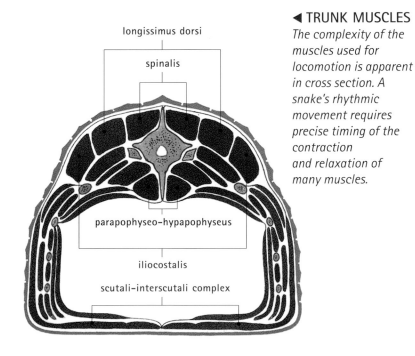

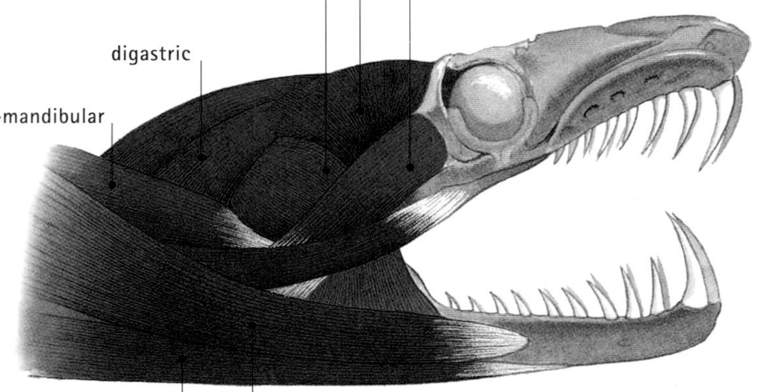

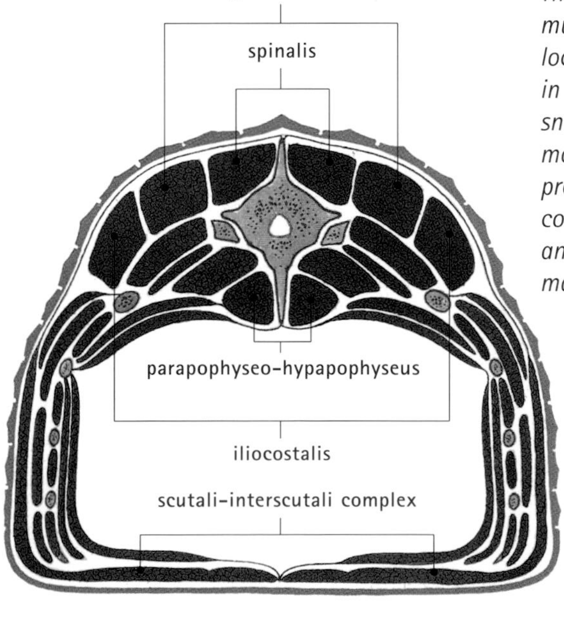

◀ TRUNK MUSCLES

The complexity of the muscles used for locomotion is apparent in cross section. A snake's rhythmic movement requires precise timing of the contraction and relaxation of many muscles.

CLOSE-UP

Shivering pythons

When a female python has laid her eggs, she coils her body around herself and around the eggs. In some species the muscles at the sides of the body then begin to rhythmically contract and relax. This is a kind of shivering, although it happens much more slowly than shivering in humans. Enough heat is generated by the contracting muscles to warm the eggs so that they have a higher temperature than their surroundings. This speeds their development. Shivering also protects the eggs from becoming chilled if the air temperature drops suddenly, as may happen, for example, in a desert at night. This kind of behavior is not seen in boas such as anacondas, because they retain their eggs internally and give birth to live young.

Nervous system

The brain, spinal cord, and peripheral nerves (those that branch from the spinal cord) of snakes show few major differences from the general vertebrate arrangement, although the nerves associated with the limbs are missing. The eyes of snakes, however, are unique. When a lizard or a bird accommodates its eye (focuses the lens to look at an object that is near or far away) the ciliary muscles surrounding the lens contract or relax. Contraction causes the lens to be squeezed. This makes the lens longer from front to back, and the front surface becomes more curved. The squeezing effect is aided by small bones called scleral ossicles that surround the iris.

Snakes do not have scleral ossicles. Instead, the ciliary muscles are at the base of the iris. When they contract, they increase the pressure in the fluid called the vitreous humor, which lies at the back of the eyeball. This increased pressure forces the lens forward and alters the focus. Unlike lizard and bird lenses, a snake's lens never changes shape.

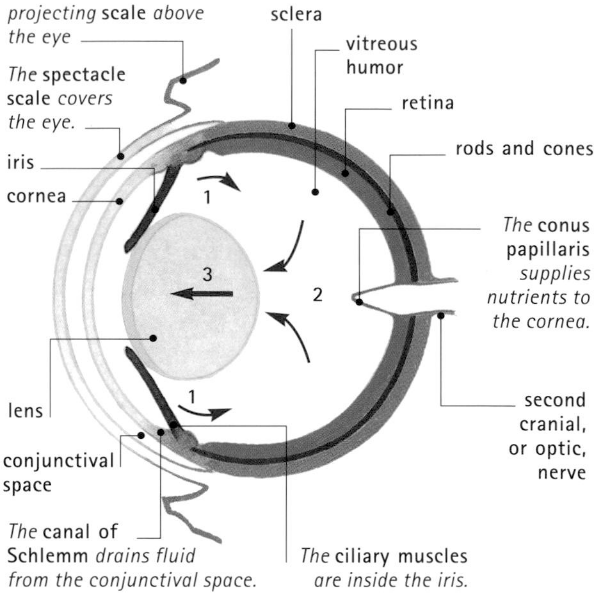

projecting scale *above the eye*
sclera
vitreous humor
The spectacle scale *covers the eye.*
retina
iris
rods and cones
cornea
1
The conus papillaris *supplies nutrients to the cornea.*
3
2
lens
1
second cranial, or optic, nerve
conjunctival space
The canal of Schlemm *drains fluid from the conjunctival space.*
The ciliary muscles *are inside the iris.*

▶ EYEBALL
Snake
Snakes have a unique focusing mechanism. The ciliary muscles contract (1), increasing the pressure on the vitreous humor (2). This moves the lens forward (3). Biologists think that snakes spent some of their evolutionary history burrowing underground, when they may have lost their sight. When snake ancestors returned to life above ground, sight re-evolved, but differently from sight in other animals.

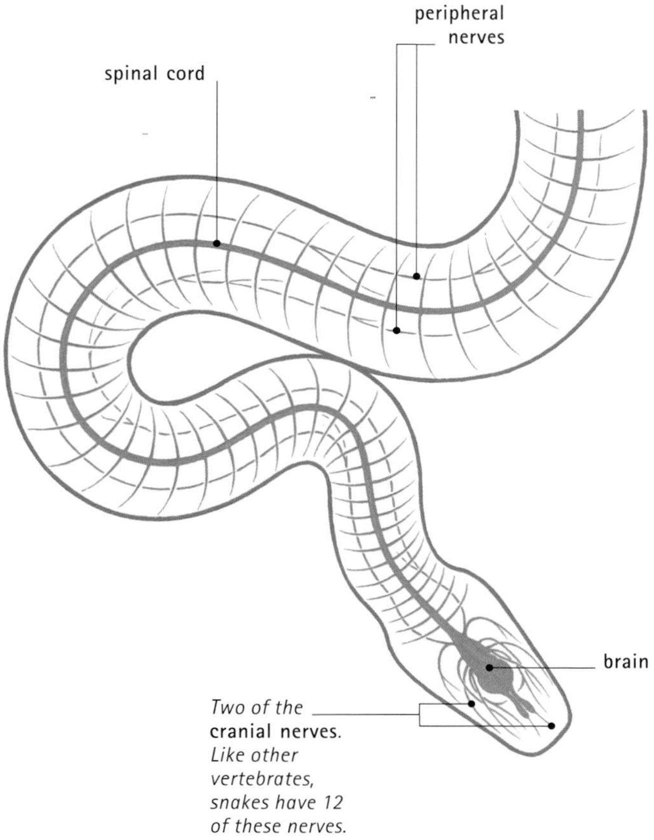

peripheral nerves
spinal cord
brain
Two of the cranial nerves. *Like other vertebrates, snakes have 12 of these nerves.*

◀ Green anaconda
An overview of a green anaconda's nervous system. Peripheral nerves repeat along the length of the body.

Seeing heat

Vipers, rattlesnakes, pythons, and boas are able to detect infrared (heat) radiation. This is especially useful for snakes that hunt by night, because it allows them to "see" warm-blooded prey in the dark. Vipers and rattlesnakes have pit organs on each side of the head. These are small depressions between the eyes and the nostrils; the cells that detect heat radiation are located inside. Many pythons and boas have a number of temperature-sensitive pits between the scales of the upper jaw. In green anacondas these pits are too small to be seen by the naked eye.

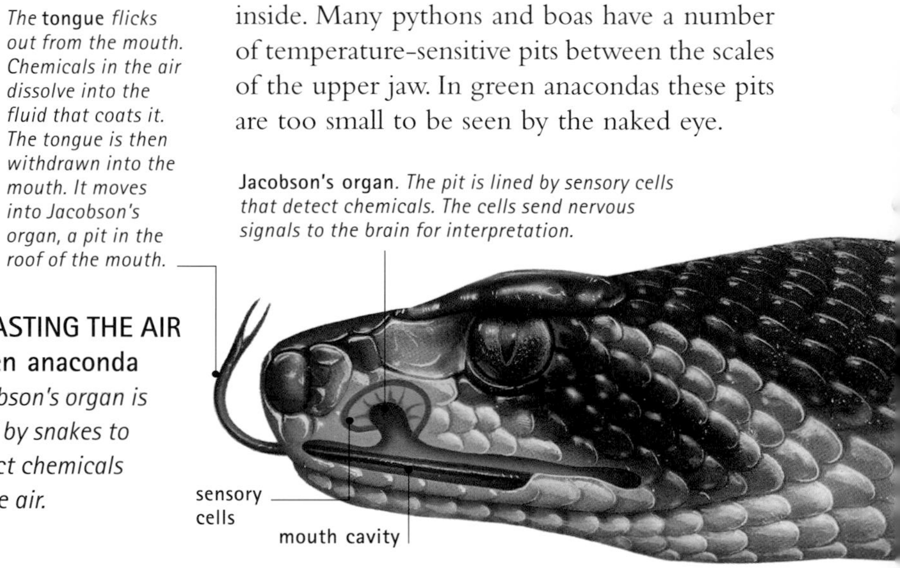

The tongue *flicks out from the mouth. Chemicals in the air dissolve into the fluid that coats it. The tongue is then withdrawn into the mouth. It moves into Jacobson's organ, a pit in the roof of the mouth.*
Jacobson's organ. *The pit is lined by sensory cells that detect chemicals. The cells send nervous signals to the brain for interpretation.*
sensory cells
mouth cavity

▶ TASTING THE AIR
Green anaconda
Jacobson's organ is used by snakes to detect chemicals in the air.

Circulatory and respiratory systems

CONNECTIONS

COMPARE the anaconda's adaptations for efficient breathing with those of an ***EARTHWORM***.

COMPARE the anaconda's three-chamber heart with the heart of other reptiles, such as a ***CROCODILE***.

Snakes have a heart with three chambers. They are a left and right atrium and one ventricle. Oxygen-rich blood from the lung enters the left atrium and is pumped into the ventricle. Since the openings of the systemic arteries (arteries that supply the organism with oxygenated blood) are on the left side of the ventricle, most of the oxygenated blood then passes into them. They take the blood to the tissues of the body. Having given up its oxygen to the tissues, blood returns to the right atrium through veins. The right atrium pumps the blood into the ventricle. This deoxygenated blood enters on the right side of the ventricle; most passes to the pulmonary artery, which is also on the right side, and along it to the lung.

This three-chamber heart is not as efficient as the four-chamber heart of birds and mammals. That is because there is some mixing of oxygen-rich and oxygen-depleted blood in the ventricle. The heart of a snake is slightly elongated from front to back compared with other reptile hearts. This enables the heart to fit into the long, thin body. In all snakes, the left systemic artery (the first major branch from the aorta) is larger than the right. This is the opposite of the condition in other reptiles and in amphibians.

One-lung wonders

Most snakes have only a single lung for breathing. In anacondas, as in most snakes, it is the right lung that has been retained. The germ of a left lung appears during embryonic development, but it does not develop very much. The right lung is long and thin and extends about one-third of the length of the body. Most of the interior of the lung has a spongy appearance because of the many small air passages it contains. They provide a large surface area for exchanging oxygen and carbon dioxide. The rear part of the lung, however, has few of these air passages and is essentially a hollow sac. The transport of air into and out of the lungs is mainly driven by movements of the ribs.

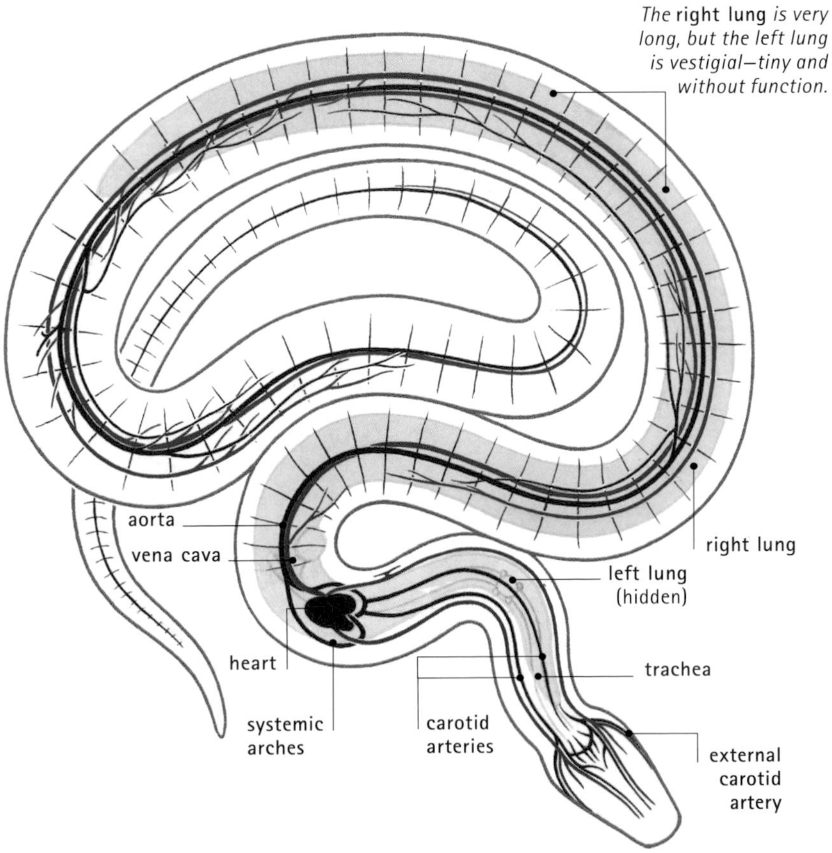

▼ **Green anaconda**
Note the size of the right lung. It extends for almost one-third of the entire length of the animal.

COMPARATIVE ANATOMY

Losing a lung

Many of the internal organs of long, thin reptiles and amphibians are themselves elongated. This allows them to fit into the confined body space without being squashed. Whereas most amphibians, reptiles, birds, and mammals have two lungs, in many elongate animals one of the lungs is much smaller than the other. In snakes the right lung is always much larger than the left, although the difference in size varies from species to species. This is also the case in elongate limbless lizards and in caecilians (legless, wormlike tropical amphibians). In one group of long reptiles, the worm-lizards or amphisbaenians, however, it is the left lung that is very much larger than the right.

Digestive and excretory systems

The most striking feature of an anaconda's digestive system is the massive stomach. Like most of the internal organs of snakes, it is long and thin, and extends for about a quarter of the length of the body. Both the stomach and the esophagus, which connects the stomach to the mouth cavity, can be greatly expanded. This enables them to accommodate enormous prey items: a large anaconda can swallow a deer or a pig, animals that have a greater diameter than an anaconda's body.

Once the prey is in the stomach, digestion begins. The prey starts to disintegrate at this stage, so the rest of the digestive system—the small and large intestines and the rectum—does not need to be able to expand. The last part of the digestive system is a series of chambers called the cloaca. Ducts bringing waste from the kidneys open into the cloaca, and the two hemipenises in males are near its base. The two penises pop out through an opening called the vent at the rear of the male's body when the animal copulates.

Removing waste

The kidneys of snakes are long and slender. They do not lie side by side. Instead, they overlap, with the right kidney starting about two-thirds of the way along the left kidney. Each kidney has a separate duct, the ureter, which opens into the cloaca.

The main excretory product in snakes is uric acid, which is voided in the form of a white paste. The elimination of uric acid requires very little water. Snakes do not need to store large amounts of urine, as mammals do, and they do not have a urinary bladder. This helps snakes conserve water.

▲ *Digestion in action. The prey of this green anaconda has been pushed down into the great snake's stomach.*

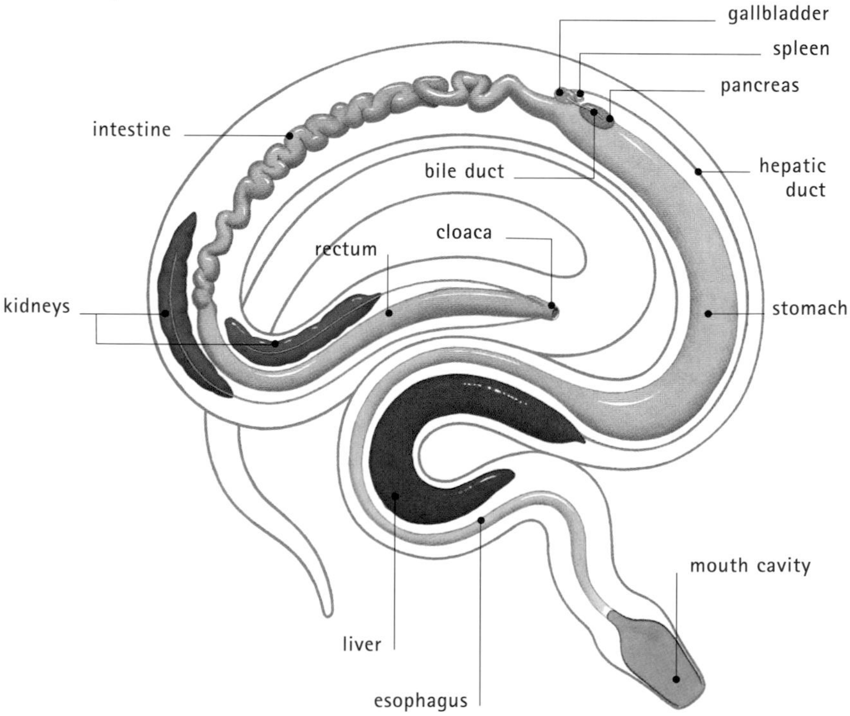

▼ *The stomach and esophagus of the green anaconda can expand enormously.*

CLOSE-UP

Feeding frequency

Snakes such as anacondas, which eat very large prey, take a long time to digest it. A meal of such a size, however, provides a lot of energy, and anacondas need to eat only infrequently. Large anacondas probably feed only once every three to four weeks. Since each meal provides so much energy, anacondas can live a long time without feeding. Large individuals can survive for more than a year without a meal. This ability is unique to snakes—no other vertebrates can live for as long without food.

Reproductive system

The testes of male anacondas lie in front of the kidneys; the right testis is further forward in the body than the left. Sperm from each testis is transferred to the cloaca along a convoluted duct called the vas deferens. Each vas deferens opens into the back part of the cloaca. This is also where the two penises (or hemipenises) occur. For most of the time the hemipenises are housed in muscular sheaths, but during copulation one of them is pushed out through a hole called the vent. From there it enters the vent of the female.

The ovaries of female anacondas lie in positions equivalent to the testes of the males. Each ovary opens into the cloaca via an oviduct. Unlike many other snakes, female anacondas do not lay eggs but retain them in the oviducts until they hatch. These snakes then give birth to fully formed young.

IN FOCUS

Eggs or live young?

Most of the 2,700 species of snakes lay eggs, but in many species the eggs hatch inside the female; fully formed young are born instead. This is called ovoviviparity. In some snake groups, such as the boas, all species produce fully formed young. In others, some lay eggs and some do not. In a few cases, even very closely related species may differ. The southern smooth snake in Europe produces fully formed young, for example, whereas the northern smooth snake lays eggs. Ovoviviparity has both advantages and disadvantages. The main advantage is that since the eggs are inside the female's body, she has some control over their incubation. She can seek out warm places to speed their development. However, an ovoviviparous snake must carry a heavy clutch for much longer than an egg-laying (or oviparous) one. A heavy clutch slows the female down and makes her vulnerable to predators.

How mating happens

Green anacondas mate in April through May. Mating usually takes place underwater. A "breeding aggregate" often forms. This is a ball consisting of one adult female intertwined with up to 12 of the much smaller adult males. Females attract the males by releasing chemical attractants into the water. A breeding aggregate may persist for as long as four weeks, but most last for a shorter time than this. In a single breeding aggregate several males may mate with the same female and thus fertilize eggs within the same brood.

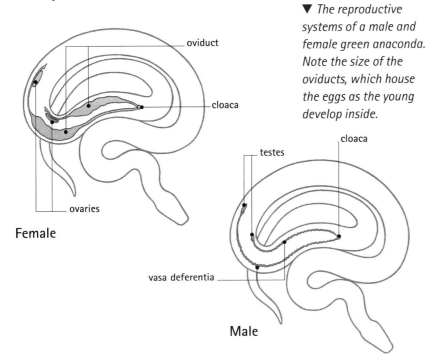

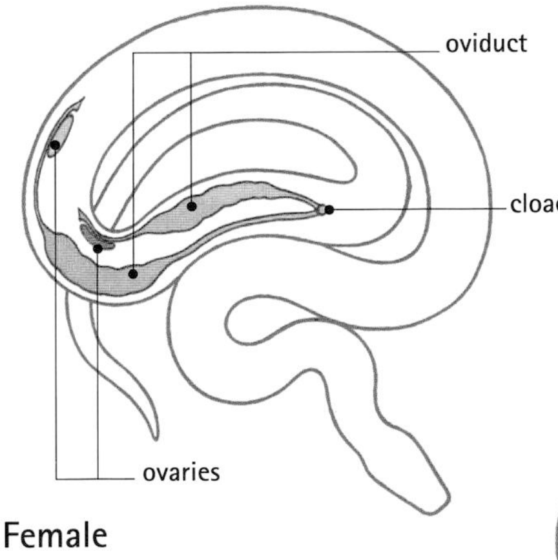

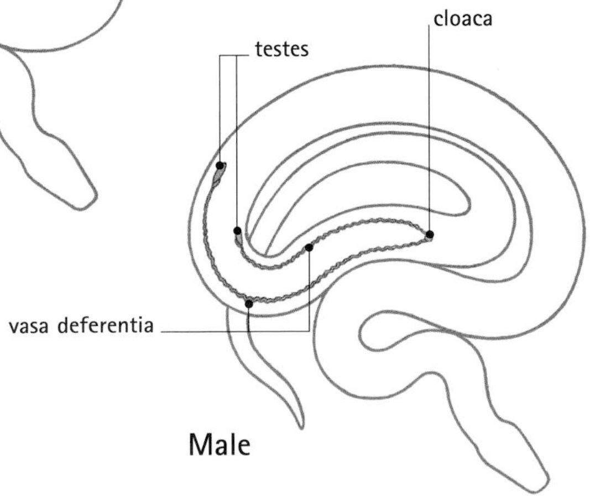

▼ *The reproductive systems of a male and female green anaconda. Note the size of the oviducts, which house the eggs as the young develop inside.*

Giving birth

Young are born in October or November, also underwater. There are usually between 20 and 40 baby snakes, but a very large anaconda may give birth to as many as 100. The young measure between 28 and 32 inches (70 and 80 cm) long at birth. They grow rapidly and reach sexual maturity at three to four years of age.

ROGER AVERY

FURTHER READING AND RESEARCH

Mattison, Christopher. 2003. *Snakes of the World.* Facts On File: New York.

Murphy, John C., and Robert W. Henderson, 1997. *Tales of Giant Snakes: A Historical Natural History of Anacondas and Pythons.* Krieger Publishing Co.: Melbourne, FL.

Grizzly bear

ORDER: Carnivora FAMILY: Ursidae
SPECIES: *Ursus arctos* SUBSPECIES *horribilis*

Grizzly bears are a subspecies of the brown bear, *Ursus arctos*. Grizzlies live in the cold forests of the Rocky Mountains and across western Canada and Alaska. Other brown bear subspecies live throughout the Northern Hemisphere.

Anatomy and taxonomy

Scientists classify all organisms into taxonomic groups based largely on anatomical features. Grizzlies and other brown bears belong to the Ursidae or bear family, which also includes black, sun, sloth, spectacled, and polar bears.

- **Animals** All animals are multicellular (many-celled). They get the energy and materials they need to survive by consuming other organisms. Unlike plants, fungi, and the members of other kingdoms, animals are able to move around for at least one phase of their lives.

- **Chordates** For at least some of their lives, all members of the Chordata have a stiff, internal rod called a notochord running along their back.

- **Vertebrates** In vertebrates, the notochord develops into a backbone, or spine, made up of small units called vertebrae. As well as a backbone, a vertebrate's body is also supported by an internal skeleton made from bone and cartilage. Paired sets of muscles enable the body to move.

- **Mammals** Mammals are a large group of warm-blooded animals. Unique features include hairs covering the body, milk glands in females to feed young, and a lower jaw formed by a single bone that hinges directly to the skull.

- **Placental mammals** Unlike marsupials, the other main mammal group, placental mammals nourish their developing young while they are still inside the mother via an organ called the placenta. The placenta is a temporary structure inside the uterus that connects unborn young to the mother's blood supply.

- **Carnivores** The word *carnivore* is often used to mean any animal that eats meat, but it is also the correct taxonomic name of a large group of placental mammals. As their name

▶ *Land carnivores are divided into two great lineages: the doglike Canoidea and the catlike Feloidea (not shown). Bears are doglike carnivores. The position of the giant panda is uncertain; some biologists place it with the bears (as in this article), but others disagree and classify this species among the procyonids (raccoons and relatives) or in a separate group.*

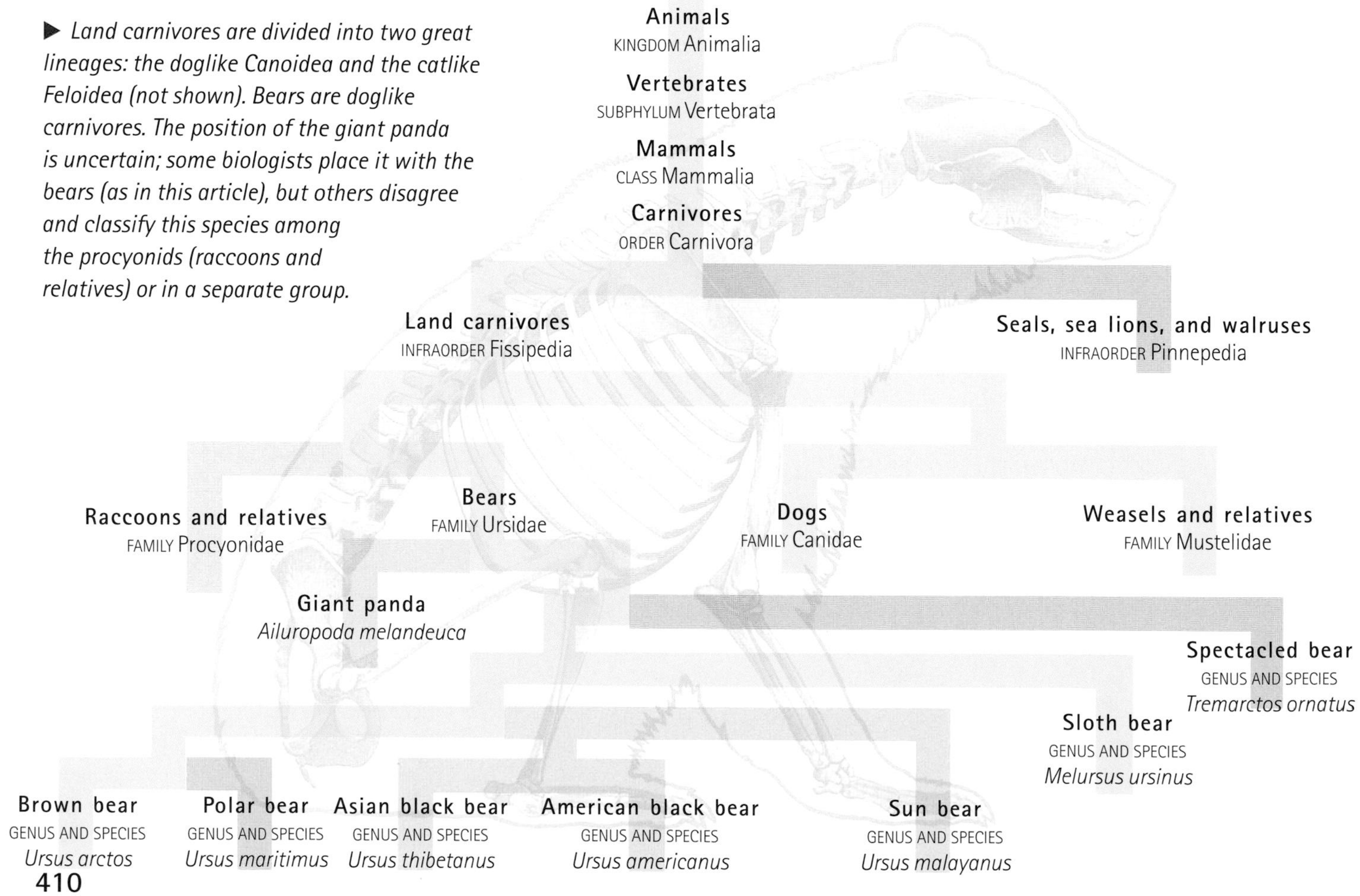

suggests, most are meat eaters and are equipped with sharp teeth and claws. A unique carnivore characteristic is that two pairs of cheek teeth, the carnassials, are shaped to cut up flesh with a scissorlike action. Carnivores include large hunters such as cats, seals, dogs, and hyenas, and smaller animals such as civets, mongooses, and weasels.

• **Bears** These are the largest carnivores, although they do not hunt as often as most other members of the order. There are eight species of bears. All have a large head with a long snout and powerful jaws. Bears have a sturdy and barrel-shaped body, and their legs are short but powerful. Bears have large, flat paws armed with long nonretractable claws. The varied diet of bears is reflected in their teeth. Their carnassials are flattened so they do not slice food but grind it up. This is useful for animals that eat a lot of plants.

• **Giant panda** The giant panda is an unusual bear (and is thought by some biologists to belong to a different group altogether). It has distinctive black and white fur, with black ears and oval spectacles around its eyes. Pandas have a unique sixth digit on their forepaws. This bony extension sticks out from the wrist. It acts like a thumb, allowing a panda to hold bamboo shoots, its main food.

• **Black bears** There are two species of black bears, the American black bear and the Asian black bear. Both are small bears that collect much of their food from trees. They rip off bark and gouge out grubs with their short claws. They also have flexible lips that help them pluck fruits. Black bears have very strong hind legs. This makes them the most adept of all bears at walking on two legs.

• **Polar bear** These are among the largest living land carnivores. They have a pale cream coat that helps them blend into icy Arctic landscapes. Their thick fur helps them stay warm, and a layer of blubber under the skin also keeps out the cold. A polar bear's outer guard hairs are hollow and translucent. These hairs trap heat from the sun and carry it down to the bear's black skin, where it is absorbed. Hairs on the paw pads help reduce the loss of precious body heat through the feet.

▲ *A foraging grizzly bear. Note the massive hump over the animal's shoulders, a characteristic unique to brown bears.*

• **Brown bears** There are several subspecies of brown bears around the world, generally distinguished by their fur color and body size. The grizzly is the main subspecies in North America, although it is now rare outside Canada and Alaska. Its brown coat is flecked with silver hairs. Kodiak bears, the largest of all the subspecies, live along the southern coast of Alaska. Eurasian brown bears are much smaller. Several Eurasian subspecies are close to extinction, such as the mazaalai from the Gobi Desert of Central Asia.

FEATURED SYSTEMS

EXTERNAL ANATOMY Grizzly bears are large carnivores with a muscular hump behind the shoulder. They have a dense brown coat which is flecked, or grizzled, with silver and gray hairs. *See pages* 412–415.

SKELETAL SYSTEM The grizzly bear's skeleton is sturdy since it must support a large and very powerful body. *See pages* 416–419.

MUSCULAR SYSTEM A grizzly's muscles are positioned to produce immense strength rather than speed of movement. *See pages* 420–421.

NERVOUS SYSTEM Grizzly bears need to have a good memory so they can remember where food sources are located. They have an excellent sense of smell, although they cannot see or hear particularly well. *See pages* 422–423.

CIRCULATORY AND RESPIRATORY SYSTEMS Although they are generally slow-moving, grizzlies are capable of sudden bursts of speed, with the large heart and lungs supplying oxygen to the muscles. *See pages* 424–425.

DIGESTIVE AND EXCRETORY SYSTEMS Being carnivores, bears have a short digestive system suited to breaking down meat. Therefore grizzlies have difficulty digesting the plant material they often eat. *See pages* 426–427.

REPRODUCTIVE SYSTEM Grizzlies are born after about eight months of gestation. They are blind, almost hairless, and completely helpless at birth. *See pages* 428–429.

External anatomy

Grizzly bears are among the largest land carnivores. Everything about them is big. They have a broad, massive head and a stocky body. Male grizzlies can reach more than 7 feet (2 m) long and weigh 1,200 pounds (0.5 metric tons) or more. Females are about the same length but are generally about half the weight of the males.

Silvertips

Grizzly bears have a thick coat of hair, which helps them keep warm in their cold habitat. The coat, or pelage, is made up of different hair types. Longer guard hairs form a shaggy protective covering. A thicker growth of fine fur grows beneath the guard hairs. The fur and the guard hairs trap a layer of air close to the skin. Since air is a poor conductor of heat, this layer prevents the bear's body heat from escaping to the surroundings. Grizzly bears are named for the color of their guard hairs. The guard hairs of other brown bear

9 feet
(2.8 m)

Adult grizzlies can be huge, reaching 9 feet (2.8 m) when standing on their back legs. Kodiak bears, a separate brown bear subspecies from coastal Alaska, are even larger.

The **fur** *varies greatly in color from near-white through brown and black. The tips of most hairs are lighter, giving them a silver-flecked (or grizzled) appearance. The coat is long and shaggy in winter, keeping the bears warm as they hibernate. It is molted in spring to reveal a shorter summer coat.*
The **body** *is barrel-shaped and immensely powerful. Coastal brown bears tend to be larger than inland grizzlies. This is because they have regular access to rich supplies of protein in the form of fish.*
The **tail** *(not visible) is very short and serves little function.*
The powerful and muscular **hind legs** *provide a bear with surprising speed; bears can outrun humans and most other animals over short distances.*
The **soles** *are covered by tough, relatively hairless pads of wrinkled skin.*

CLOSE-UP

Bear hair

The long guard hairs that form a grizzly's outer coat are tipped with silver and gray, but the shorter fur beneath is unusual for another reason. Most mammal hair has an "agouti" pattern. This is a series of light and dark rings, which together give the hair its overall color. The hairs in bear fur, however, are unusual because they have just a single color throughout. The color varies from brown and red to pale yellow. Humans are one of the few other types of mammals to have single-color hairs. Hair color is produced by proteins called melanins. The agouti pattern is produced by two different types of melanins. Eumelanins are very dark and pheomelanins are light. Agouti-type hairs contain both types of melanin, whereas bear fur contains just a single type.

The large **dorsal hump** distinguishes brown bears from other species. The hump contains muscles that add power to the forelimbs.
Grizzlies have good hearing, but the **external ears** are small and furry. This helps minimize heat loss.
The **head** is wide, containing powerful muscles that drive the jaw muscles.
Bear **eyes** are small relative to the rest of the head. Their visual powers are broadly similar to those of humans. Unlike some other carnivores, bears have color vision. This helps them identify ripe fruits and nuts.
Bears have an excellent sense of smell. This helps them find buried food and detect carrion over long distances.
The **lips** are separate from the gums and are tremendously flexible.
At more than 4 inches (10 cm) long, the **front claws** are longer than those of any other bear. The grizzly uses them for digging out dens, scratching at bark, or slashing at prey or rival bears. Unlike black bears, grizzlies do not use their claws for climbing.
The **front paws** are vital food-gathering tools. They are used for fishing, digging up roots or small rodents, or lashing out at prey.

◀ A grizzly bear's external anatomy. A grizzly's mass varies greatly over the course of a year; fat is deposited throughout summer and fall but is used up as the animal hibernates.

CONNECTIONS

COMPARE the claws of a sloth bear with those of other specialist social insect-eaters, such as a ***GIANT ANTEATER***.

COMPARE the paws of a climbing bear, such as a black bear, with those of other expert climbers such as a ***CHIMPANZEE***.

subspecies do not change color along their length. Many of a grizzly's guard hairs, however, are tipped with silver or gray. This has the effect of making the coat look flecked, or grizzled, with silver. Grizzlies are also called silvertips for this reason.

Big head

In proportion to its body, a grizzly bear's head is very large when compared with that of most other animals. Unlike most other bears, grizzlies have a distinctive concave face, which curves up from their long snout into a wide forehead. Their eyes are small when compared with the wide, rounded head. They sit in the middle of the face on either side of the snout.

The bear's ears are small and heavily furred. Small ears help the animal retain its body heat. Larger ears would radiate more heat into the air. Grizzly bears' ears are often obscured by the long, shaggy coat, especially during the winter months. The animal's hair is longer at this time to help it conserve heat.

CLOSE-UP

Grizzly feet

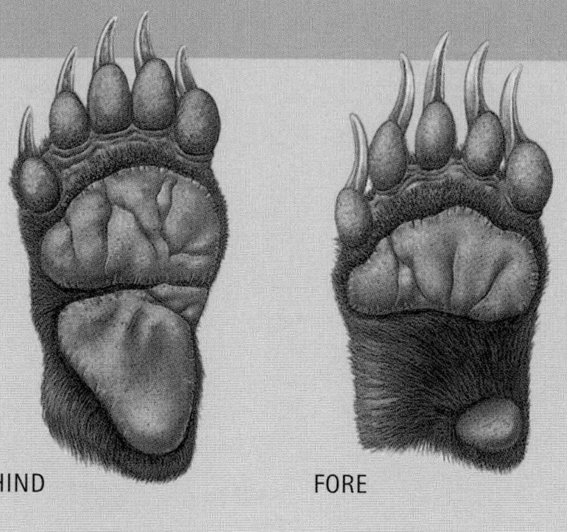

A grizzly's foreclaws are long and strong, and have a range of uses, from digging to fishing.

Grizzlies' feet are equipped with huge claws, longer than those of any other bear. The claws on the forefeet measure around 4 inches (10 cm) long, longer than those on the hind feet. The feet themselves may be up to 16 inches (41 cm) in length.

Unlike most cats, bears cannot retract their claws into a protective sheath when not in use. The claws are therefore subject to a great deal of wear and tear. An older bear with several worn or broken claws may struggle to dig up food during lean periods or to make a den for hibernation in the fall.

Although grizzlies' forefeet are very dexterous and capable of plucking fruits and leaves, the long claws make it harder for them to grip trunks and climb trees. The claws of American black bears are much shorter than those of grizzlies. These smaller bears are expert climbers that spend a lot of time resting or foraging in trees.

Like other bears, grizzlies have a long snout with a pair of large, hairless nostrils. Uniquely among carnivores, bears do not have whiskers—touch-sensitive hairs that project from the snout. They do, however, have an extremely good sense of smell. Hair does grow on the thin upper lip between the nose and mouth. A grizzly's lips are mobile and not attached to the gums. This makes them ideal for scooping up small berries or insects.

CLOSE-UP

A large hump

One distinguishing feature of grizzly bears is the pronounced hump between the shoulders. This is a huge mass of muscles, which helps make the bear's forelimbs enormously strong. The strong limbs are used to club prey or rival bears, to dig out a den, and to dig up roots and other buried foods.

Silhouettes of a grizzly bear (left) and an American black bear (right).

▼ PAWS OF SMALLER BEARS

Paw shape and structure give some clues about the lifestyle and feeding habits of some of the smaller species of bears.

Sloth bear

Sloth bears feed extensively on termites. Their front claws are long and the forelimbs powerful to help them rip into the insects' nests. Sloth bears also have protrusible lips, and their front incisor teeth are missing. Thus the lips can form a tube through which the termites can be sucked.

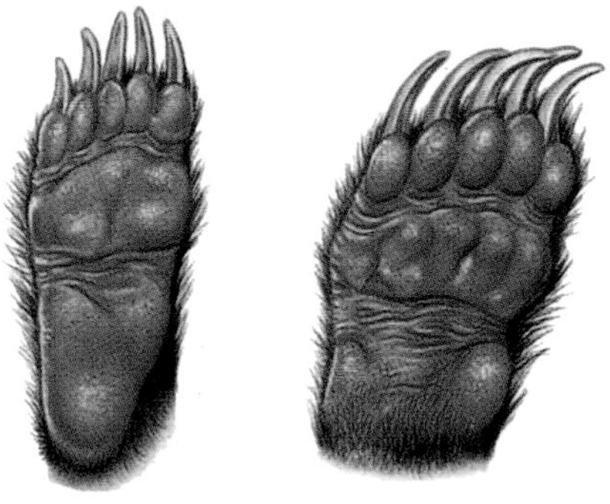

Sun bear

Sun bears eat a range of foods, such as small mammals, the growing tips of palm trees, and termites. The claws help the bears climb. They often clamber up trees in search of bees' nests, which they break into with their front claws to get at the honey inside.

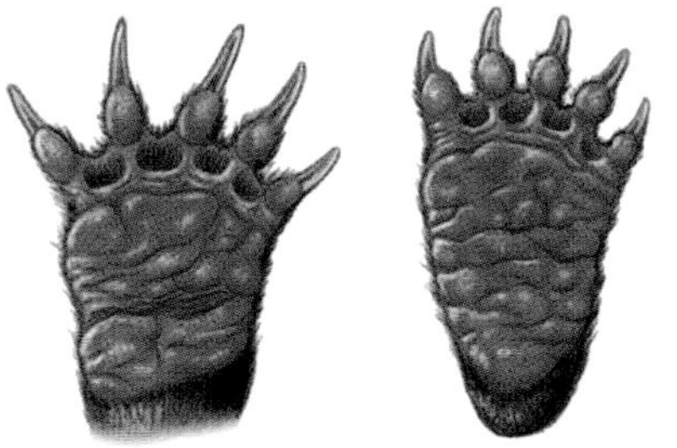

Asian black bear

Black bears eat a lot of plant material, especially in the fall when berries and nuts are available. They use their claws to overturn logs in search of tasty grubs. Fish, small mammals, and the young of larger animals are also eaten by black bears. Black bears spend much of their time in the trees.

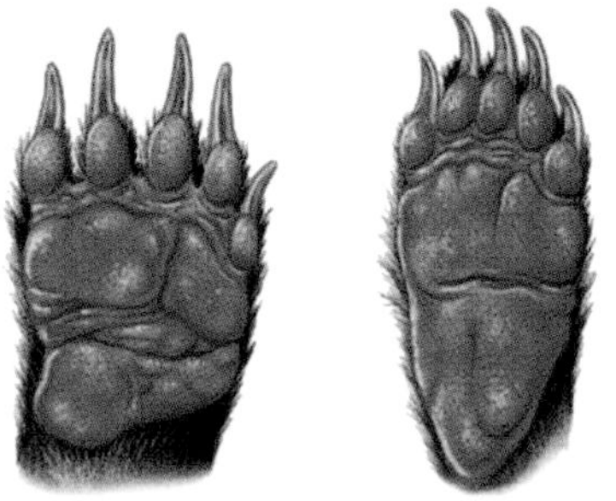

American black bear
Black bears have a good sense of smell, which helps them find birds' nests and carrion, but it is less important than in other bears, since much of their food is plant material. Their muzzle is relatively short for a bear.

Brown bear
Brown bears have a very good sense of smell and a longer muzzle than a black bear. They eat a wide range of foods and will feed on carrion when they can find it. Smell helps them find hidden foods such as buried roots and tubers.

Polar bear
Polar bears have a long muzzle. This provides for an extremely sensitive sense of smell, essential in a barren habitat where food is always scarce. A polar bear can smell a carcass from as far as 20 miles (32 km) away.

▲ **MUZZLE SHAPE**
The size and shape of a bear's muzzle relate to its feeding habits, habitat, and lifestyle.

A strong body

Although grizzlies can stand on their hind legs, they are generally quadrapedal (they usually move on four legs). The legs are short but very thick, ending with wide, paddlelike paws. Unlike most other bears, grizzly paws have little hair on the soles. Grizzly bears have a distinctive hump between their shoulders, where the huge sets of muscles that power the forelegs and neck are anchored. However, the dorsal hump is not the tallest point on the bear's body. The lumbar region of the spine, just in front of the pelvis, curves up to form the tallest point along the back.

Unlike most other carnivores, bears have only a short tail. The tail appears to be vestigial, a structure left over from a long-tailed bear ancestor that no longer has a function. The lack of a tail is unusual for an animal that often climbs trees, as do most bears apart from grizzlies; in climbing, a long tail might help the animal with balance.

EVOLUTION

From small beginnings

The first bears appeared around 20 million years ago. The earliest known specimen, the dawn bear *Ursavus*, dates from around this time. Unlike modern bears, the dawn bear had a long bushy tail and a small, agile body similar to that of a raccoon. Modern bears separated into three main groups around 10 million to 12 million years ago. One group is represented today by the giant panda; another, the running or tremarctine bears, by the spectacled bear from South America. The third and largest group, the ursines, includes grizzlies and all other living bears.

Grizzlies and polar bears are the largest modern land carnivores, but they would be dwarfed by an ancient tremarctine called *Arctodus*, the giant short-faced bear. This massive beast lived in North and South America until as recently as 12,000 years ago. It weighed more than 1,800 pounds (820 kg) and could reach 11 feet (3.4 m) tall on its hind legs. Running bears like *Arctodus* had longer legs than grizzlies; they could chase down quick prey including bison, horses, and camels.

The giant short-faced bear was a powerful pursuit predator. Note how long its legs were compared with those of a grizzly bear.

Skeletal system

CONNECTIONS

COMPARE the plantigrade stance of a bear with the digitigrade stance of other carnivores, such as a ***WOLF***.

COMPARE the grinding carnassial teeth of a brown bear, which is largely omnivorous, with the shearing carnassials of a carnivore that eats mostly meat, such as a ***LION*** or ***PUMA***.

Grizzly bears have the same basic skeleton as other types of carnivores, but one that is adapted for supporting their huge, powerful body. Other carnivores have a lighter skeleton built for speed at the expense of strength.

Short and sturdy

While faster-running and more agile types of carnivores have long legs with thin bones, grizzlies and other bears have relatively short legs for their overall body size. The bones inside the legs are able to support the great weight of a grizzly's body because they are extremely thick. Large bears, such as grizzlies and polar bears, are much heavier than other large terrestrial carnivores. For example, tigers, the largest of the cats, may grow to the same length as a grizzly—or occasionally even longer—but they weigh only about half as much. Bears' bones, therefore, have to be much thicker to support the extra weight.

Bears' limbs are also straighter than those of most carnivores. Animals that need to run fast or make long leaps to catch prey tend to have limbs that are held in a bent position at the knee and elbow. This allows the legs to act like springs. The joints are flexed as the animal lands after a bound. They then spring back, releasing energy that is used to propel the animal into the next bound. A bear's legs are not kept bent in this way, since they would collapse under its weight. One effect is that, although capable of bursts of speed, modern bears are not efficient runners and soon tire. As a result, bears tend to rely on ambush when hunting larger mammals.

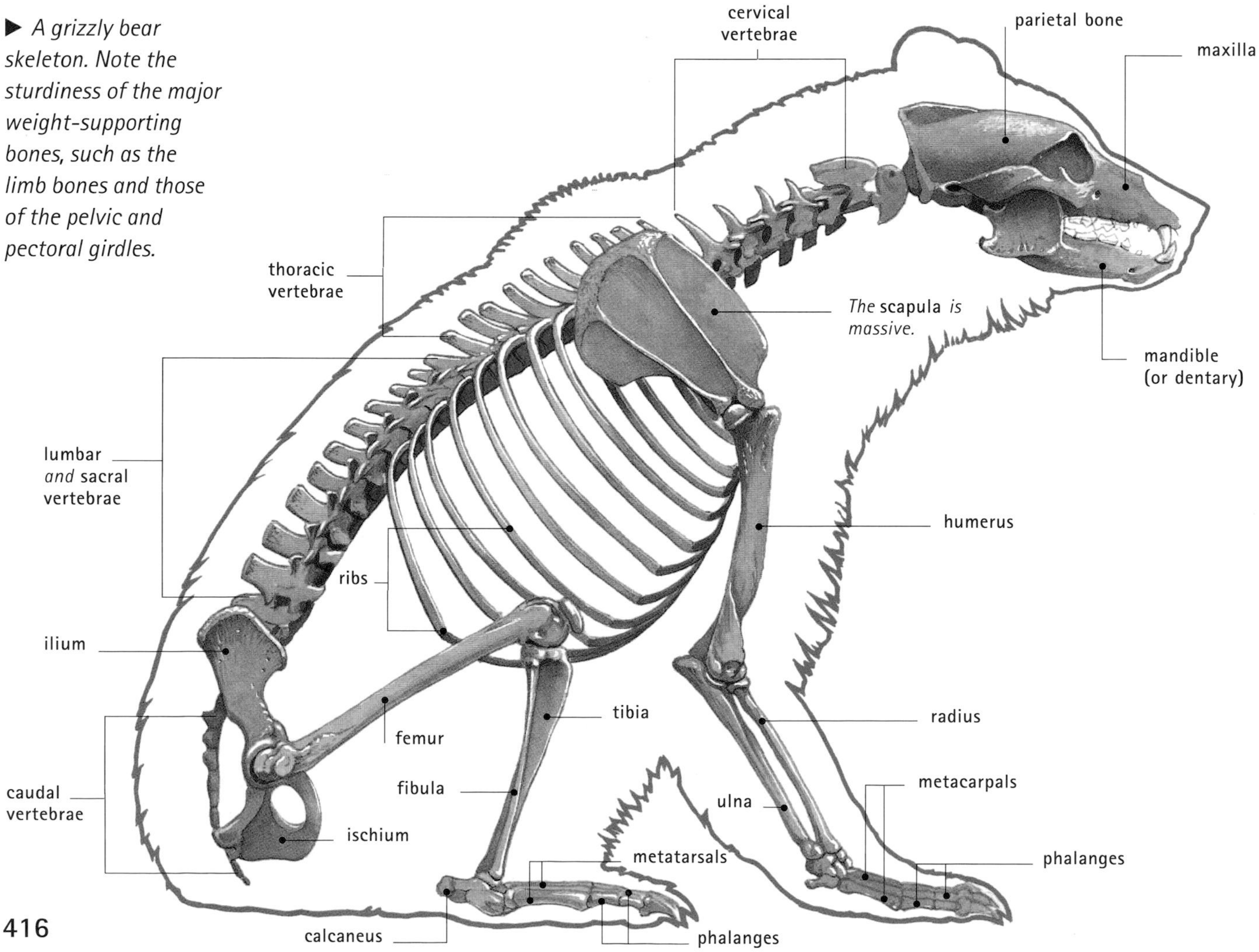

► *A grizzly bear skeleton. Note the sturdiness of the major weight-supporting bones, such as the limb bones and those of the pelvic and pectoral girdles.*

CLOSE-UP

Massive skulls

Grizzly bear skulls are large and elongated. The long skull supports a long, narrow snout, which is filled with many odor-sensitive cells. There is also room for several long, flat teeth inside the powerful jaw. Other carnivores need less room for the smaller teeth they use for slicing meat. Grizzlies have a huge brain case, which acts as a solid anchorage for the powerful temporalis muscles on each side of the head. These muscles raise the lower jaw, delivering the forces needed to crush tough material.

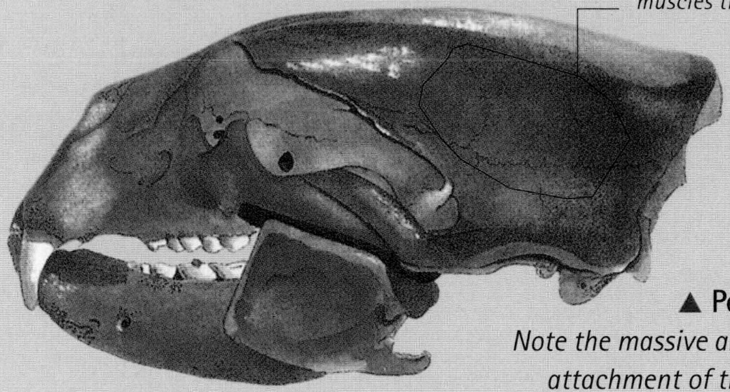

Attachment area for the main muscles that power the bite.

▲ Polar bear skull
Note the massive area available for attachment of the masseter and temporal muscles, which give the polar bear a powerful bite.

▶ Grizzly bear skull
Bones of the skull and jaws of a grizzly bear. The massive canines are used more for display than for hunting or feeding.

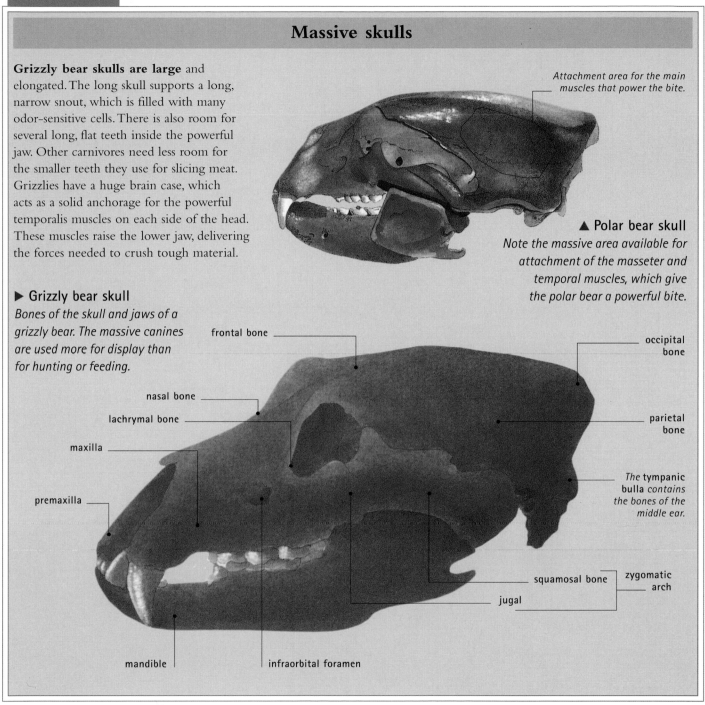

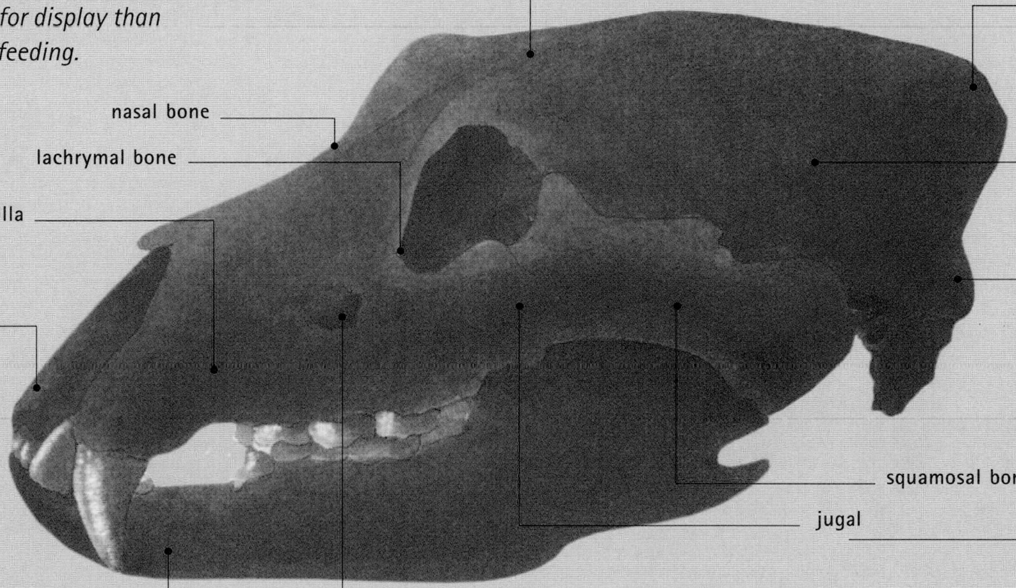

A small collar

Like other carnivores, grizzly bears have a short, very slender collarbone, or clavicle. This bone connects the scapulae (shoulder blades) to the sternum (breastbone). The clavicle is joined to these bones by long, flexible ligaments. In other mammals, including humans, the clavicles are much longer and more robust. They also have more sturdy connections with the other bones. In humans and other primates, for example, the clavicles are used to keep the scapulae in a fixed position, and they also provide an anchorage for the muscles that are used to swing the arms out to the sides. Bears and other carnivores do not need to move their forelimbs in this way. In fact, movements out to the sides would weaken their limbs and prevent them from running quickly without injuring themselves. Since the clavicle is not needed to give any structural or muscular support, it has been reduced to just a sliver of bone with little real function at all.

Most carnivores are digitigrade—they walk on their toes. Bears are different. Bears are flat-footed. When they walk, the whole length of each foot—from heel to toe—touches the ground. This arrangement is called a plantigrade stance. This is an adaptation that helps support the great weight of the animal.

Bears share with other carnivores an unusual bone in their feet. Most mammals have several bones in the wrist or ankle, but in carnivores three of these are fused, forming the scapholunate bone. It was once thought that this bone was used for running and acted as a shock absorber. It seems more likely that the bone serves as a firm anchorage for the muscles that bend the paw at the wrist, helping the animal climb or grapple with prey.

COMPARATIVE ANATOMY

Flat feet

Both humans and bears have plantigrade feet, but this way of walking is unusual for a mammal. Most carnivores, for example, are digitigrade. Digitigrade animals walk on their toes, not the soles of their feet. Such animals are generally fast runners. Standing on their toes makes their legs longer and lengthens their stride. Digitigrade animals also have relatively longer limb bones.

Grizzly bears are generally slow and steady walkers, and many of the other bear species are adept climbers. Their flat feet provide a good support for their heavy, muscular body. Being able to use the entire length of the foot to grip onto a surface helps in climbing. Many climbing rodents, such as squirrels, are also plantigrade for this reason. Plantigrade feet also make it easier for grizzlies and other bears to rise up on to their hind feet to get to hard-to-reach food or to intimidate rivals.

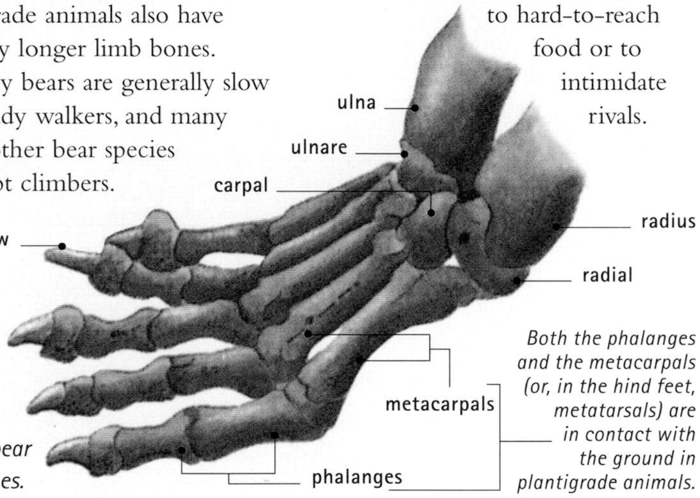

ulna
ulnare
carpal
claw
radius
radial
metacarpals
phalanges
Both the phalanges and the metacarpals (or, in the hind feet, metatarsals) are in contact with the ground in plantigrade animals.

Grizzly bear foot bones.

Grizzly teeth

Grizzlies have 42 teeth. They grow a full set by the time they are about two and a half years old. Like most other types of mammals, grizzly bears have four types of teeth: incisors, canines, premolars, and molars. An adult grizzly bear has 12 incisors at the front of the jaw: 6 in the upper jaw, and 6 in the lower. They are small but sharp and are used as slicing teeth. Behind the incisors, there are four canines, one on each side of the mouth in both the upper and the lower jaw. Canines are long, pointed fangs. Most carnivores use their canines to bite prey and to rip away chunks of flesh. Although grizzlies do use them occasionally to kill prey, their canines are unsuitably large for their

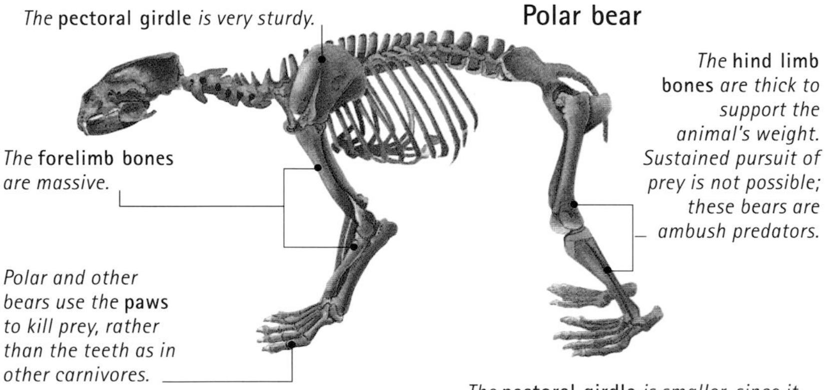

Polar bear
The pectoral girdle is very sturdy.
The hind limb bones are thick to support the animal's weight. Sustained pursuit of prey is not possible; these bears are ambush predators.
The forelimb bones are massive.
Polar and other bears use the paws to kill prey, rather than the teeth as in other carnivores.

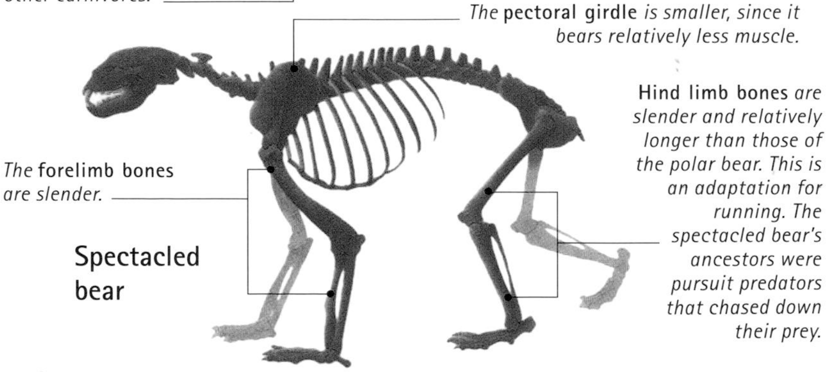

The pectoral girdle is smaller, since it bears relatively less muscle.
Hind limb bones are slender and relatively longer than those of the polar bear. This is an adaptation for running. The spectacled bear's ancestors were pursuit predators that chased down their prey.
The forelimb bones are slender.
Spectacled bear

◀ THE LAST OF THE RUNNING BEARS
The spectacled bear is a running, or tremarctine, bear. In this illustration, a spectacled bear skeleton has been scaled up to the size of a polar bear. Polar bears belong to the other bear group, the ursines. Although spectacled bears feed mainly on plant material, their skeletons show that their ancestors were more active predators. There were many other tremarctines in the past, some of which pursued and killed large mammal prey.

COMPARATIVE ANATOMY

Wolverines

A wolverine looks as if it might be related to bears or dogs, but it is in fact a type of mustelid—a relative of the weasels, otters, and skunks. However, it shares many anatomical features with bears.

A grizzly bear is much bigger than a wolverine, being about 10 times as heavy and twice as long. However, both animals have a thick, sturdy skeleton with short muscular legs. The legs end in wide, plantigrade paws. This configuration makes both animals good diggers. A wolverine's wide paws are especially useful in the winter, when deep snow makes moving around hard for larger animals, including grizzlies. A wolverine's feet act like snowshoes, preventing the animal from sinking into the snow. What wolverines lack in size, they make up for in aggression. During the winter, these creatures are remarkably fierce. Wolverines have been known to drive adult grizzlies from their prey.

Another reason why wolverines are often mistaken for dogs or small bears is their robust head and large snout. Wolverines also share with bears an acute sense of smell and extremely powerful jaws. They use their jaws to gnaw meat that has frozen solid or to drag carcasses over long distances.

▼ **LOWER JAW**

Grizzly bear

Bones and teeth of the lower jaw. Processes and condyles are extensions of bones.

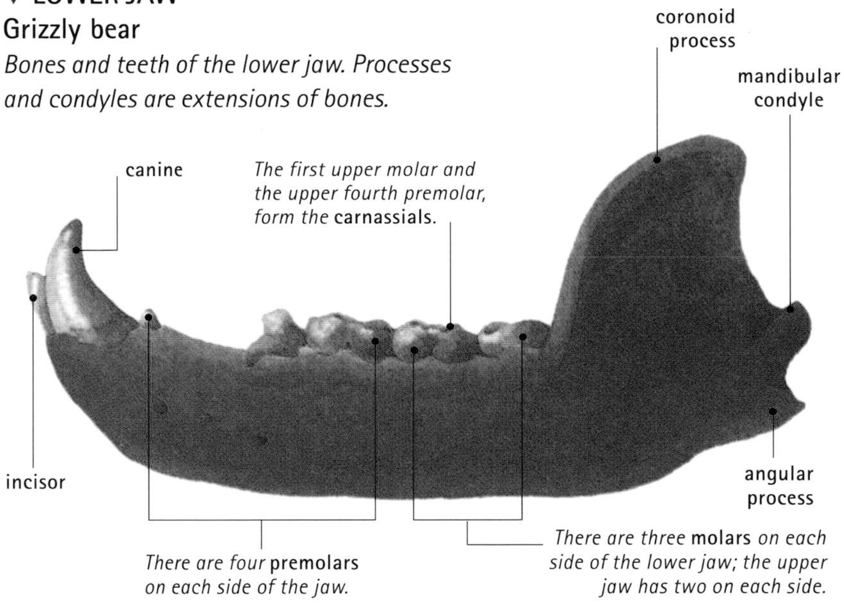
coronoid process
mandibular condyle
canine
The first upper molar and the upper fourth premolar, form the carnassials.
incisor
angular process
There are four premolars on each side of the jaw.
There are three molars on each side of the lower jaw; the upper jaw has two on each side.

general diet. However, bears display their large canines as a warning of their fighting skills when threatened by a rival.

Grizzlies have two types of cheek teeth. They have eight premolars in each jaw, and four molars in the upper jaw and six in the lower one. The lower first molars and the upper fourth premolars are sometimes called carnassials. Originally adapted for shearing through flesh and hide, they are found only in members of the Carnivora; even the giant panda, which feeds almost exclusively on bamboo leaves and shoots, has two pairs of carnassial teeth. Unlike in most other carnivores, the cheek teeth of bears, including the carnassials, are flat, like millstones, making them suitable for grinding up fibrous plant material. These grinding teeth are bunodont; that is, they have four low cusps, giving them a rough surface that is ideal for crushing up food. A cusp is a point on the surface of a tooth. By comparison, dogs have just a single cusp on their molars.

▶ **SKULL**

Sloth bear

The skull viewed from below with the lower jaw removed, showing the bones of the palate and base of the skull. Note the sloth bear's unique dental adaptations. Several foramina are shown. The word foramen *means "window." A foramen is a hole in bone through which blood vessels or nerves pass.*

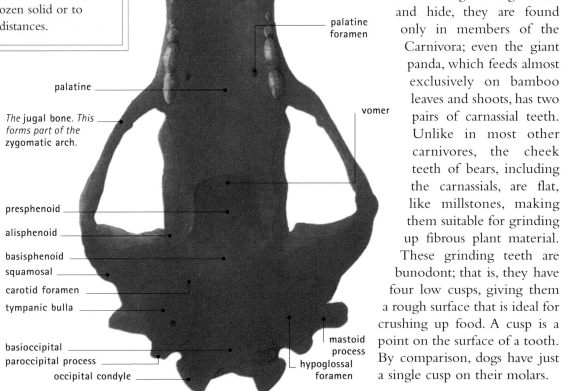

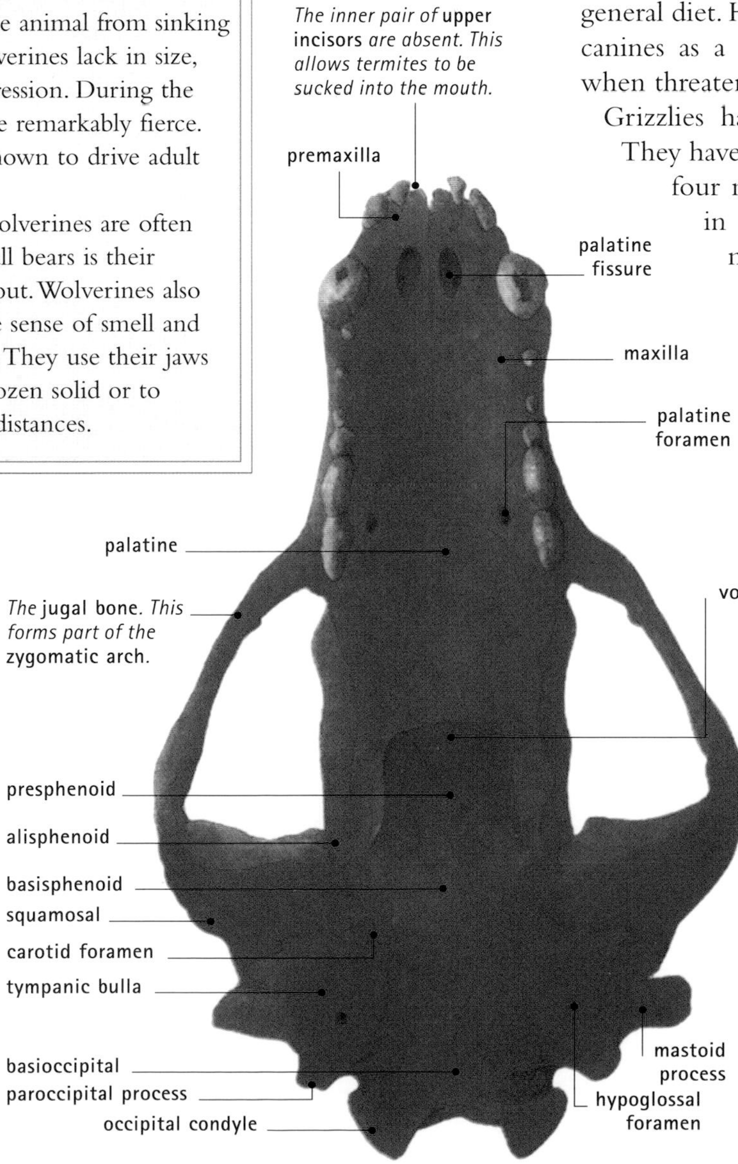
The inner pair of upper incisors are absent. This allows termites to be sucked into the mouth.
premaxilla
palatine fissure
maxilla
palatine foramen
palatine
The jugal bone. This forms part of the zygomatic arch.
vomer
presphenoid
alisphenoid
basisphenoid
squamosal
carotid foramen
tympanic bulla
basioccipital
paroccipital process
occipital condyle
mastoid process
hypoglossal foramen

Muscular system

CONNECTIONS

COMPARE the forelimb muscles of a grizzly bear with those of other carnivores. A ***WEASEL***'s forelimb muscles are relatively much smaller than a bear's, since weasels use their teeth to kill prey. A ***LION***'s are larger than a weasel's; a lion may use its forelimbs to hold or trip fleeing prey, but again uses its jaws to make the kill.

The phrase "as strong as a bear" is an apt one. No other animal of its size has strength comparable to that of a grizzly bear. The animal's strength comes in part from the rigid anchorage that the thick skeleton provides, and the position and size of powerful blocks of muscles.

Muscle systems

The muscles are attached by tendons to processes, or outgrowths, on the bones. Tendons are inelastic cords. When a muscle contracts, it shortens in length and pulls back on the tendon. The tendon transmits the force to the bone, causing it to move.

Muscles are generally arranged in pairs, one pulling in the opposite direction from the other. Muscle fibers are made of two types of proteins, which lie side by side in long filaments. When the muscle receives a signal from the central nervous system, the two filaments slide past each other, making the fiber

PREDATOR AND PREY

Super strength

Bears use their tremendous strength in a number of ways. For example, they use their powerful forelegs to locate and catch prey. The forepaws are used to dig out burrowing mammals, to roll away rocks or logs to expose insects, and to club larger prey. Over short distances, grizzlies can outrun galloping horses, running at up to 30 miles per hour (48 km/h); they are also strong swimmers.

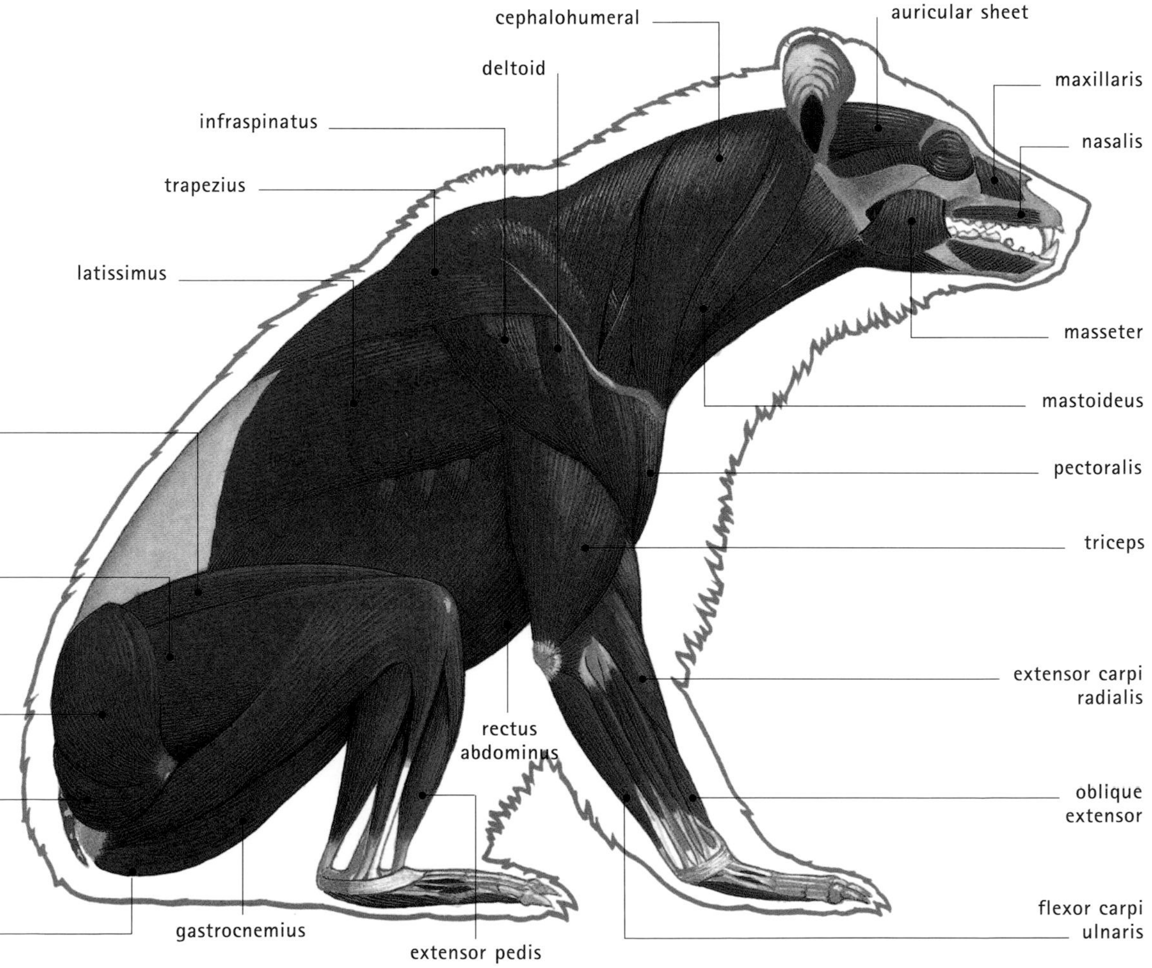

▶ *Important features of a grizzly bear's musculature. Bears are among the strongest and most powerful of all animals.*

shorten. A large muscle may contain hundreds of millions of individual filaments. Together, the filaments can create a considerable force.

Muscle positions

The bones in a limb are connected at joints. Muscles attached at each joint move the bones back and forth like a lever. The positions of the muscles affect the angle through which the bone can move, the speed at which it moves, and the power of the movement. Muscles that attach to bones very close to the joints produce a large range of movement and can make the bone move quickly. However the bone and the limb around it will not move with any great power. Therefore, the limb cannot exert a large force on another object. Fast-running animals, such as gazelles and deer, have muscle systems arranged in this fashion. The muscles of these animals deliver speed but not strength.

Muscles that attach farther away from the pivoting joint produce slower but more powerful movements. Bears' muscles are arranged this way. The processes to which they attach are also longer than in weaker but faster-moving animals. These longer bone processes increase the leverage of the muscles as they contract. Since their muscles do not attach close to joints, bears have a more limited range of movements. This inflexibility is also a result of bears being so muscle-bound.

IN FOCUS

Size matters

Although they grow to about the same length as males, female grizzly bears weigh only around half as much. Both sexes put on a lot of weight to help them survive the winter, and females nursing young tend to lose more weight than males during the winter. However, this does not explain the huge difference in body mass between the sexes.

Male bears are heavier because they are more muscular than the females. Adult bears spend most of their time living alone, but they do sometimes gather in large groups around a rich supply of food, such as a river teeming with fish. In these situations, the male bears slot into a social hierarchy, with the largest bears at the top. These heavy bears get to choose the best places to feed. They enforce their social standing by making displays of their formidable teeth, claws, and strength. On rare occasions, closely matched rivals actually fight each other, sometimes inflicting serious injury. During the breeding season, male bears also fight over access to females.

▼ FORELIMB ATTACHMENTS
How adaptations for strength and speed are reflected in the location of limb muscles and their attachments, and the dimensions of the bones themselves.

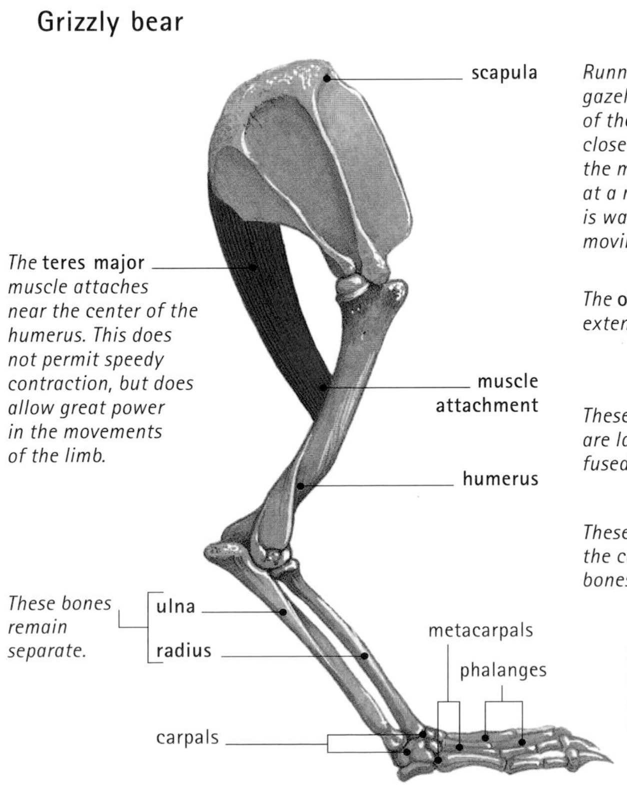

Grizzly bear
scapula
The teres major muscle attaches near the center of the humerus. This does not permit speedy contraction, but does allow great power in the movements of the limb.
muscle attachment
humerus
These bones remain separate.
ulna
radius
metacarpals
phalanges
carpals

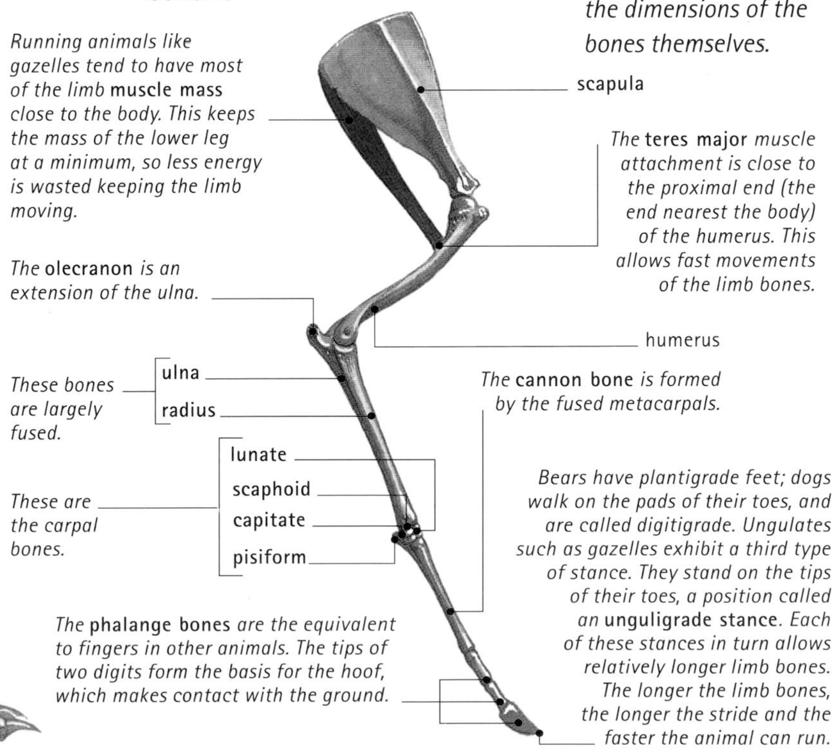

Gazelle
Running animals like gazelles tend to have most of the limb muscle mass close to the body. This keeps the mass of the lower leg at a minimum, so less energy is wasted keeping the limb moving.
scapula
The teres major muscle attachment is close to the proximal end (the end nearest the body) of the humerus. This allows fast movements of the limb bones.
The olecranon is an extension of the ulna.
humerus
These bones are largely fused.
ulna
radius
The cannon bone is formed by the fused metacarpals.
These are the carpal bones.
lunate
scaphoid
capitate
pisiform
Bears have plantigrade feet; dogs walk on the pads of their toes, and are called digitigrade. Ungulates such as gazelles exhibit a third type of stance. They stand on the tips of their toes, a position called an unguligrade stance. Each of these stances in turn allows relatively longer limb bones. The longer the limb bones, the longer the stride and the faster the animal can run.
The phalange bones are the equivalent to fingers in other animals. The tips of two digits form the basis for the hoof, which makes contact with the ground.

Nervous system

CONNECTIONS

COMPARE the memory of a bear with that of an ***ELEPHANT*** or ***CHIMPANZEE***. All animals need to remember where they can find food through the year.

COMPARE how a grizzly bear smells its surroundings with the methods used by a ***LOBSTER***. Both animals have an excellent sense of smell, but they achieve this in different ways.

The nervous system is a grizzly bear's link with the outside world. Information about the environment is collected by the animal's sense organs. The nervous system carries information from one part of the body to another in the form of electrical impulses. The central nervous system (CNS) is made up of the brain and the spinal cord. The spinal cord is a dense mass of nerve tissue that is protected by the vertebrae. Sensory information is carried to the CNS by a network of sensory neurons (nerve cells). The information is processed and signals are sent to the muscles in response. These signals travel along a separate set of nerve cells called motor neurons.

Bear senses

There are five main senses—vision, hearing, smell, taste, and touch. However, bears must also be sensitive to other environmental factors, such as temperature, light levels, and day length. A grizzly bear has an extremely acute sense of smell. The bear's long snout is lined with millions of odor-sensitive cells that can detect chemicals in the air. They form a layer called the nasal epithelia. The inner surfaces of the nose are kept moist. Chemicals dissolve into the fluid, allowing their detection. The surfaces are separated from the outside by a thick, moist pad around the nostrils. A grizzly bear relies on its nose to find food, avoid rival bears, locate mates, and identify its cubs. Male bears advertise their presence to rivals and potential mates by wiping their saliva on rocks and tree trunks.

Grizzlies can smell food, such as a rotting carcass, from several miles away. Their sense of smell is as good as that of any other mammal. Bears rely on smell more than their other senses. Like dogs, another group of carnivores

▶ *Major nerves and ganglia (nodules of nervous tissue) of a grizzly bear. Responses to information from the sense organs are generated by the central nervous system, which includes the brain and spinal cord.*

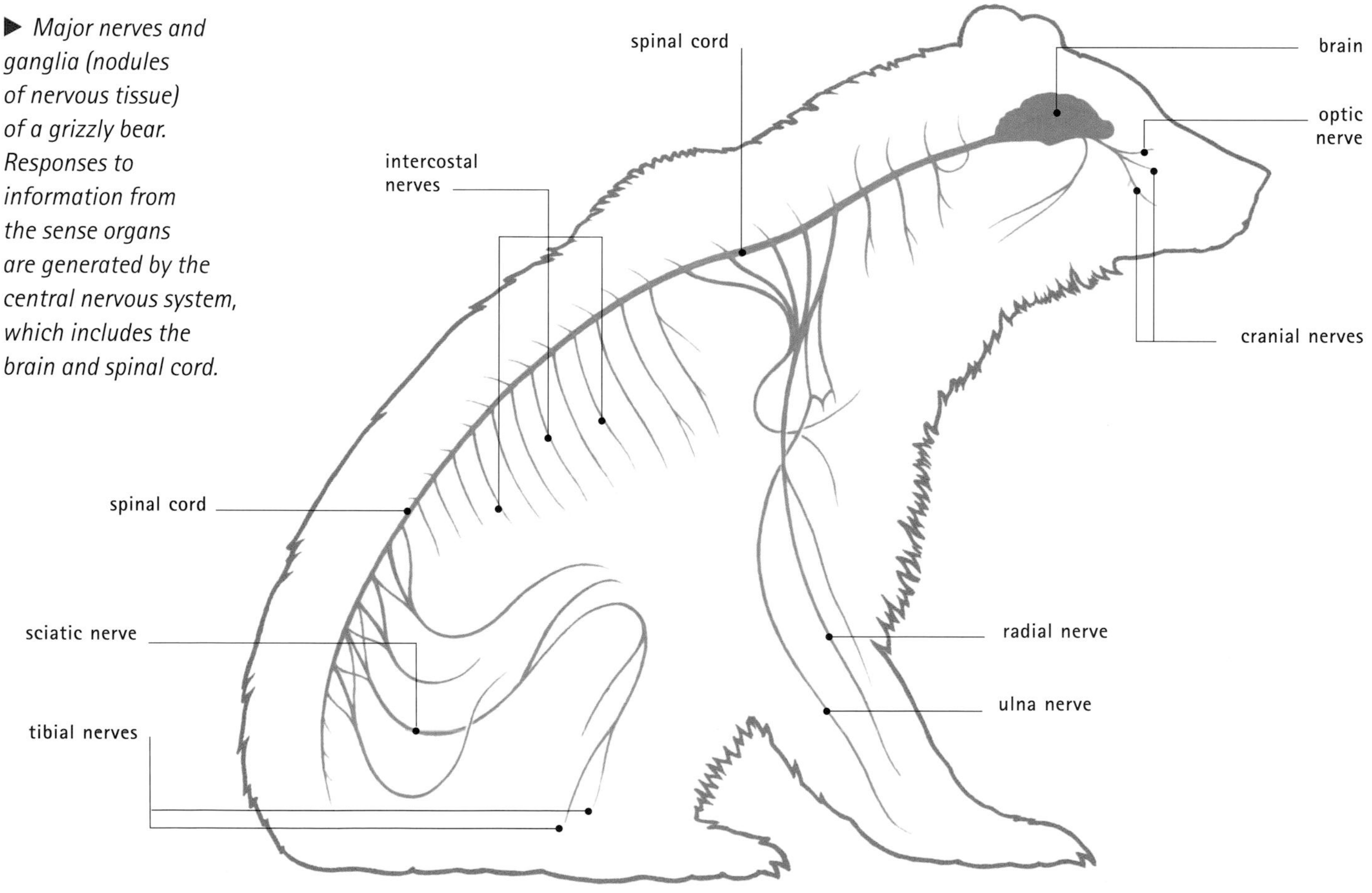

with an acute sense of smell, bears have a large olfactory lobe at the front of their brain. The olfactory lobe processes information on chemicals in the air sent by the nasal epithelial cells.

Eyes and ears

Grizzly bears have smaller eyes than humans do. They can see in color, using three types of color-detecting cone cells in the retina, which is the light-sensitive surface at the back of the eyeball. Although they do not have exceptional vision, they are able to see well in dim light, and so they can forage at dawn and dusk. They do this using a different set of retinal cells called rods. Rods are not sensitive to color, but are much more sensitive to light than cones are. The idea that bears have poor eyesight may come from the fact that they often get very close to an object before they react to it. This may not necessarily be because they cannot see the object. Rather, it may be because they prefer to smell things before acting.

A grizzly's hearing is about as acute as a human's, although grizzlies can probably hear slightly higher frequencies than most people. Hearing is especially useful for tracking small prey animals in dense forest undergrowth and for locating the position of burrows and tunnels where prey may be living.

IN FOCUS

A good memory

Grizzly bears eat a wide variety of things, many of which are available only at certain times of the year and only in certain parts of their home range. They also store food in underground caches, which they may return to months later. To do this, grizzlies need a good memory and must be able to build up a detailed mental map of their surroundings. They often follow the same well-trodden trails through their home range as they travel from one food source to the next.

IN FOCUS

Points of view

Grizzly bears are sometimes seen standing on their hind legs. This allows them visually to survey a wider area. They may also sniff the air to locate the source of an interesting odor. People often mistake bears' rising up on two legs as an act of aggression. All they are doing, however, is taking a good look around.

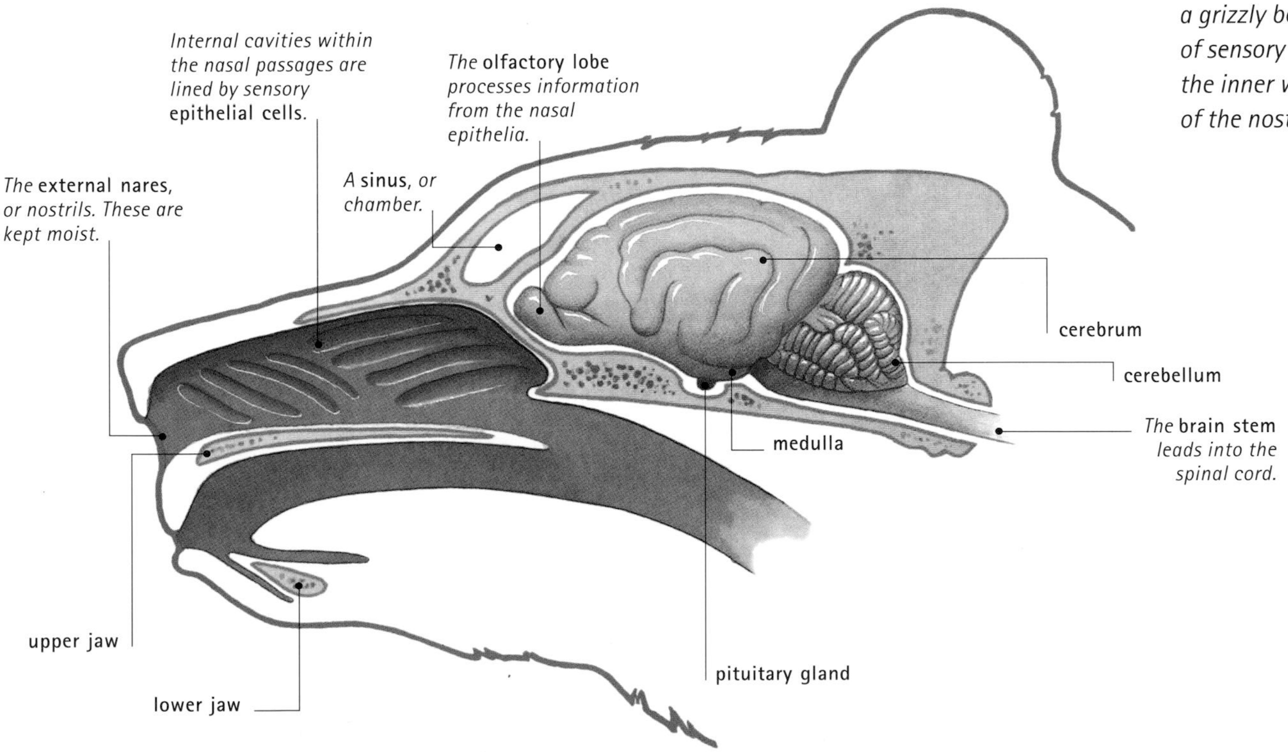

Internal cavities within the nasal passages are lined by sensory epithelial cells.
The olfactory lobe processes information from the nasal epithelia.
The external nares, or nostrils. These are kept moist.
A sinus, or chamber.
cerebrum
cerebellum
medulla
The brain stem leads into the spinal cord.
upper jaw
lower jaw
pituitary gland

▼ **BRAIN AND NOSTRIL**
A powerful sense of smell is essential for a grizzly bear. Millions of sensory cells line the inner walls of the nostrils.

Circulatory and respiratory systems

Animals need to take in oxygen to survive. The gas is absorbed into the body from the air through baglike organs called lungs. Once inside the body, the oxygen is carried around the body by the circulatory system to places where it is necessary. The circulatory system consists of tubes that carry blood—a mixture of cells and liquid. Oxygen is carried by red blood cells, which give blood its characteristic color. Oxygen is used by body cells to obtain energy from food. A by-product of this process is carbon dioxide. This waste gas is also carried by the blood until it can be removed at the lungs.

Breathing

A grizzly has large lungs, enabling it to take in the large quantities of oxygen necessary to power its big, muscular body. Air is inhaled through the nose or mouth and travels into the lungs through the windpipe, or trachea. A grizzly bear takes between 6 and 10 breaths every minute, but this may rise to 45 breaths per minute when the bear is running.

The trachea connects to a branching network of pipes that fills both lungs. The air travels into this network and fills tiny sacs called alveoli. They lie at the tips of the smallest tubes. There, the oxygen dissolves into the moisture that lines the lungs.

The alveoli are surrounded by tiny blood vessels called capillaries. The dissolved oxygen travels across the capillary walls and into the blood. There, it combines with a chemical in the red blood cells called hemoglobin. Hemoglobin also carries away some of the

CONNECTIONS

COMPARE the grizzly bear's heart and breathing rates with those of a ***HUMAN.***

COMPARE the ways that a bear regulates its temperature with those used by a ***GRAY WHALE*** and a ***WOLF.***

▶ CIRCULATORY SYSTEM
Note the very large heart, which is able to pump blood around the body quickly—and so allow lots of oxygen to be passed swiftly to the tissues of the grizzly bear.

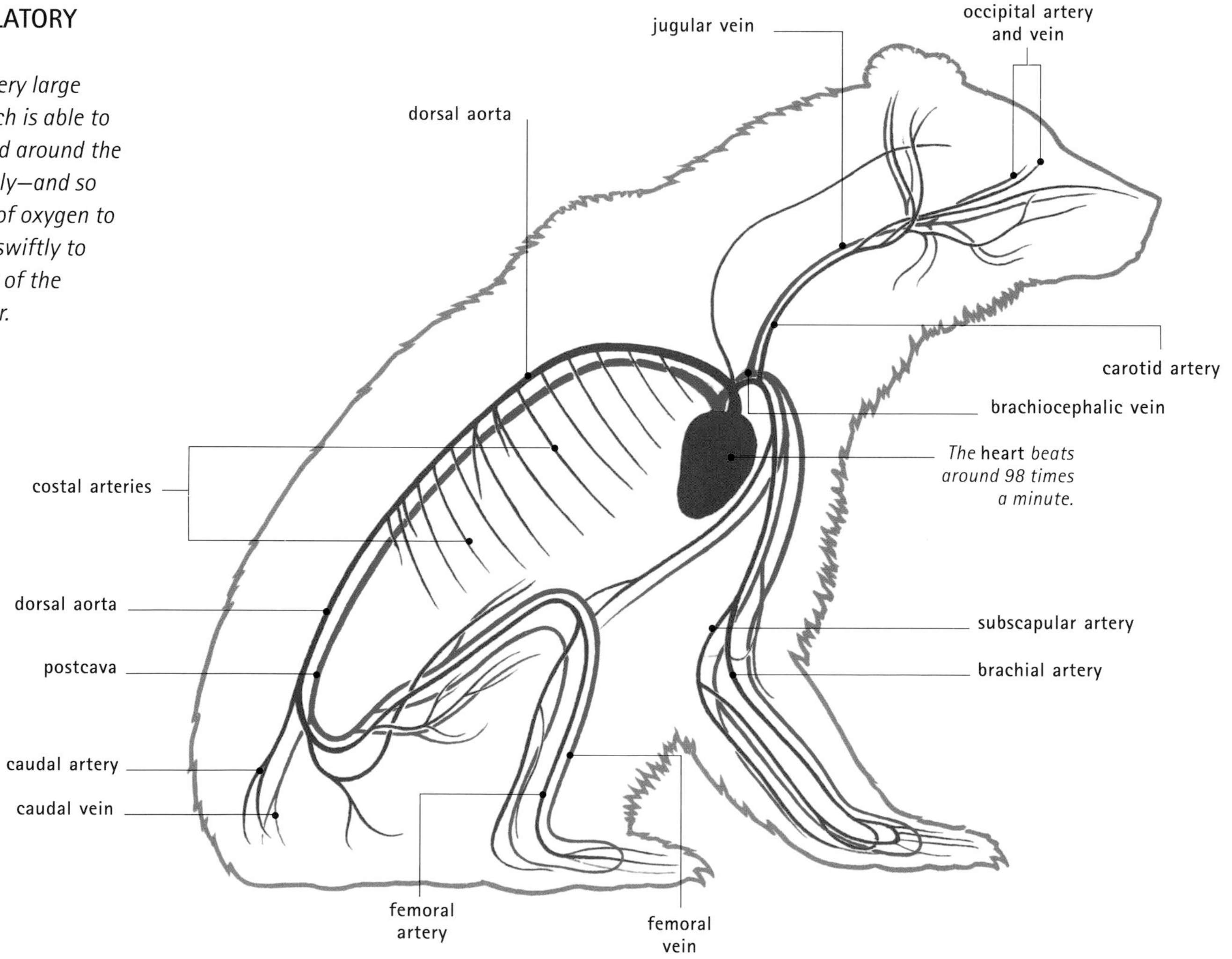

IN FOCUS

Hibernation

Grizzly bears rely on a lot of fruits and other plant foods to survive. However, in winter there is very little of this food around, and the bears are forced to become dormant, or hibernate. By the fall, a grizzly bear has put on a lot of fat, enabling the animal to survive through the winter. This need to store fat is one reason why grizzlies and some other bears are so large. In some places, grizzlies hibernate for six months of the year. The bears use their powerful forepaws to dig dens in which to hibernate. During hibernation, the bears enter a sleeplike state. Their heart rate drops to about 10 beats a minute, and they breathe only 3 or 4 times a minute.

Unlike other hibernating animals, which may experience a hibernating temperature close to 35°F (2°C), the bear's body temperature falls only to around 93°F (34°C). For this reason, some zoologists do not consider this to be true hibernation of the type practiced by animals such as marmots, bats, and dormice.

carbon dioxide that is released at the lungs. The carbon dioxide passes in the opposite direction from the oxygen, and is breathed out.

The blood system

As well as oxygen, blood carries everything else that the body needs, such as food, messenger chemicals called hormones, and white blood cells to fight invading organisms.

CLOSE-UP

Regulating temperature

Bears control their body temperature, or thermoregulate, in a number of ways. Normally a bear's body temperature is about 100°F (38°C). Fur and a fat layer keep these animals warm in cold periods, but grizzlies may become too hot during warmer weather or after periods of exertion. Bears do not sweat, but they are able to pant like dogs to lose heat. Panting helps water evaporate from the mouth, taking heat with it.

Grizzlies may also bathe in water or snow to keep cool. They often dig shallow pits in which to rest during the hottest part of the day. Their characteristic dorsal hump also comes in useful as a radiator. The muscles in the hump are full of blood vessels. Heat from the blood is radiated into the air around the hump, keeping the bear cool.

Blood is pumped by the heart. Like all mammals, grizzly bears have a single heart close to the center of the chest. The heart is a muscular pump. With each contraction, the heart forces blood along. A grizzly's heart beats around 98 times a minute when it is awake. This rate halves when the animal is asleep.

Oxygen-depleted blood from the body arrives at the right side of the heart. One of the heart chambers, the right ventricle, pumps this blood to the lungs. There it becomes enriched with oxygen before returning to the left side. This contains a more muscular pump, the left ventricle, which forces the oxygen-rich blood around the body.

▼ *This female polar bear has recently left her den in the snow. Her cubs were born as she hibernated. Male and non-nursing polar bears do not hibernate.*

Digestive and excretory systems

Grizzly bears eat a wide range of foods, often feeding on different food types at different times of the year. In spring, for example, bears eat grasses, roots, and mosses, but as the year progresses they eat more fungi, bulbs, and tubers. Whenever possible, grizzly bears catch and eat animals, anything from ants to moose. They also feed on carrion. Grizzlies often use their strong paws to dig out marmots and other burrowing rodents, and to catch salmon that are migrating up rivers from the ocean. Grizzlies are more carnivorous than other types of brown bears, although they still eat more plant material than meat.

CLOSE-UP

Scats

Bear feces are called scats. Biologists who study feces are called scatologists. Scatologists can tell a lot about how a bear lives from its scats. Since much of the food is only partially digested, they can tell what the bear has been eating. This lets them know where the bear was feeding and when it might have passed a particular location.

To disperse their seeds, many plants rely on bears and other animals to eat their fruits, such as berries. Since grizzlies can only partially digest much of their food, the seeds pass safely through the gut. They then emerge along with the scats. A fresh scat provides the germinating seeds with fertilizer, giving the seedlings a good start in life.

Short guts

Even though grizzly bears eat a lot of plant food, their digestive system is more suited to dealing with meat. Just like their carnivore relatives that eat flesh exclusively, bears have a short digestive tract. Consequently, their food passes through the body quickly. This is not a problem when bears are digesting meals containing meat and other proteins that can be broken down easily by enzymes. Plant material, however, is much harder to digest and needs to be processed for far longer than meat. Plant-eating animals such as cattle have very long, large guts. Food may take days to digest inside these types of animals. The most easily digested plant foods are ripe fruits; grizzlies target these above all other foodstuffs.

▼ *The digestive system of a grizzly bear. Note how short the gut is compared with that of a ruminant such as a giraffe.*

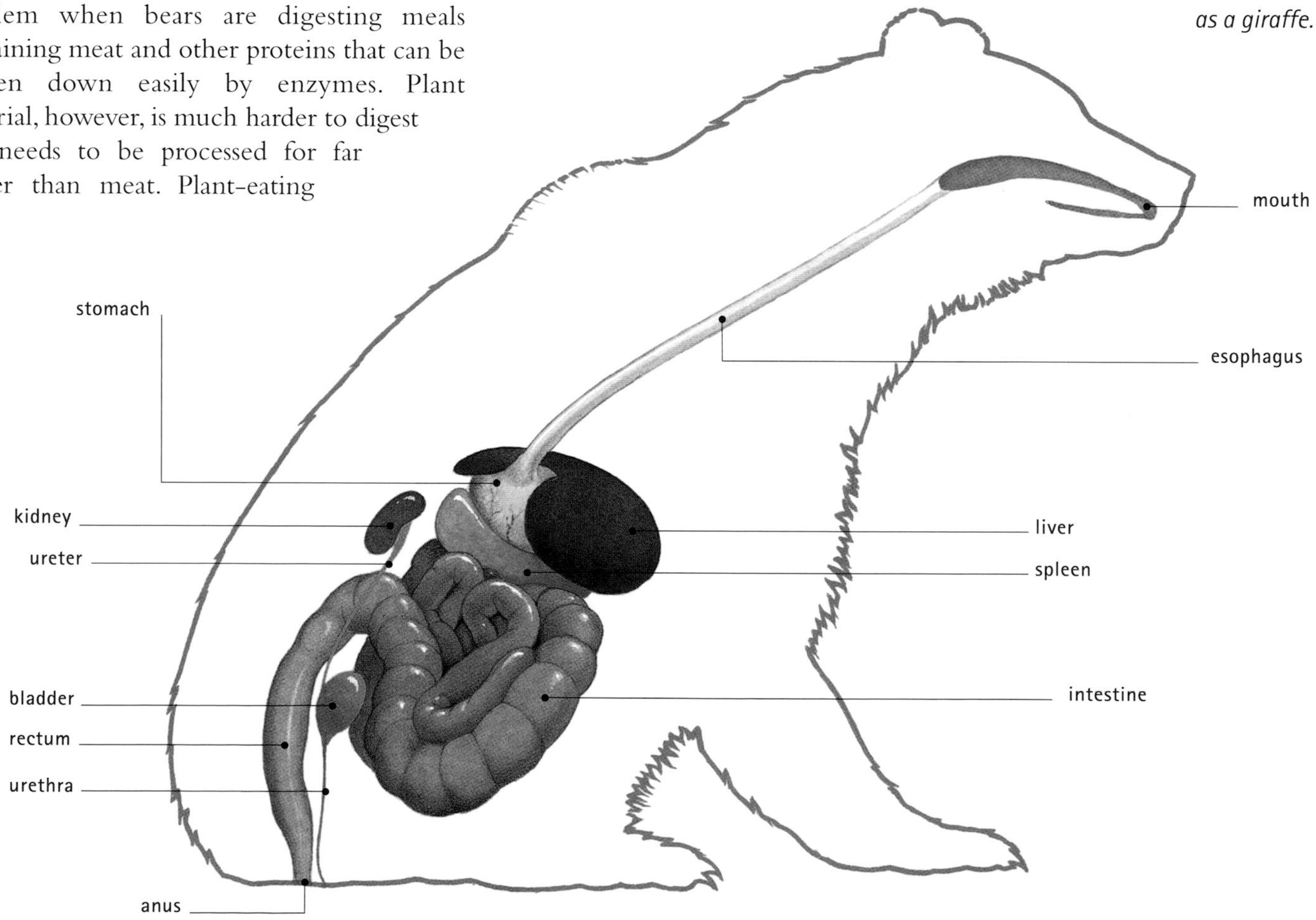

▲ *The gap between the large canines and the premolars is called a diastema. Bears pull plant stems through the diastema, ripping away leaves and buds.*

CONNECTIONS

COMPARE the grizzly bear's digestive system with that of a ruminant such as a ***GIRAFFE***. A giraffe's digestive system is much more efficient for drawing nutrients from plant leaves.

Wasting food

Since bears have a meat-eater's short digestive system but a mixed diet, they have difficulty extracting all the nutrients they need from their food. Fruits and other plant materials will sometimes pass right through the bear and remain almost completely intact.

Most digested food is absorbed into the blood through the walls of part of the small intestine, the ileum. Grizzlies and other bears have a longer ileum than other carnivores, so they can absorb more of the digested materials that pass through.

Fermentation

Tough plant fibers are made of cellulose, a carbohydrate composed of chains of sugar molecules linked together. This composition makes cellulose extremely tough to break down. Many animals rely on microorganisms that live in their gut to produce enzymes that attack cellulose. The gut microorganisms ferment the plant material, breaking it into sugars that can be used both by themselves and by the animal they live in. Grizzly bears do not do this on the scale of animals that eat only plant material. However, bears do have an extended colon, part of the large intestine, where some fermentation takes place. The large colon gives the bear its barrel-shaped body.

Excretion

Excretion is the removal of waste materials from inside the body. Most waste leaves a grizzly's body in urine. This is a liquid that contains mainly urea. Urea is a nitrogen-containing compound produced by the breakdown of proteins. It is toxic in large quantities, so it must be removed. This task is done by the kidneys.

Like all mammals, grizzly bears have two kidneys. The kidneys filter the blood to remove urea and any excess water, salts, or sugars the blood might contain. The liquid portion of the blood passes through seivelike structures around the outside of the kidney and into fine tubes. The kidney then reabsorbs things the body needs from the filtered liquid. They pass back into the blood. The remaining liquid dribbles into the bladder and is passed from the body as urine.

IN FOCUS

Recycling materials

Grizzly bears may hibernate for a period of up to six months, and while doing so they do not eat, drink, urinate, or defecate. In place of food, the bears get energy from their stores of fat. If they have laid down enough fat before winter begins, they will not lose much muscle during hibernation. Instead, the proteins the bears need to maintain their body functions are made by recycling the waste urea, which is usually excreted in urine. Fluids are also recycled from the bladder. This system keeps the bear's bones and muscles healthy through many months of inactivity each year.

Reproductive system

Like most carnivores, male grizzly bears have a baculum, or penis bone. The function of this bone is not fully understood. It may be that the bone hooks onto the female during mating. This prevents the female from breaking away from the male easily and makes copulation last longer. Longer copulation not only makes it easier for the male to deposit plenty of sperm inside the female but may also have the effect of causing her to produce eggs, or to ovulate. A short copulation might not be enough to ensure that this happens.

The mating season

Mating takes place in summer. Male and female grizzly bears seek each other out, principally by scent. A male guards his mate aggressively for around three weeks after copulation. This prevents other males from attempting to mate with her. After this time, the female bear is sure to have ovulated, so it will be the guarding male's sperm that has fertilized her eggs, not that of a rival. In many placental mammals, the fertilized eggs, or blastocysts, implant into the wall of the uterus within a few weeks. However, grizzly bear eggs are not implanted until the fall. This delayed implantation is necessary to prevent the cubs from being born too early.

Sleeping mothers

Grizzly bears give birth between January and March, producing up to four cubs. During this period, adult grizzly bears are hibernating in underground dens. Females give birth to their cubs while in this state of dozy dormancy. Grizzly bear cubs are born undeveloped and helpless. Their eyes are closed, and they have only a fine covering of fur. They generally weigh between 12 and 24 ounces

▶ *Females delay implantation of fertilized eggs into the wall of the uterus from spring, when they mate, to fall. This ensures that cubs are born at just the right time, during the winter.*

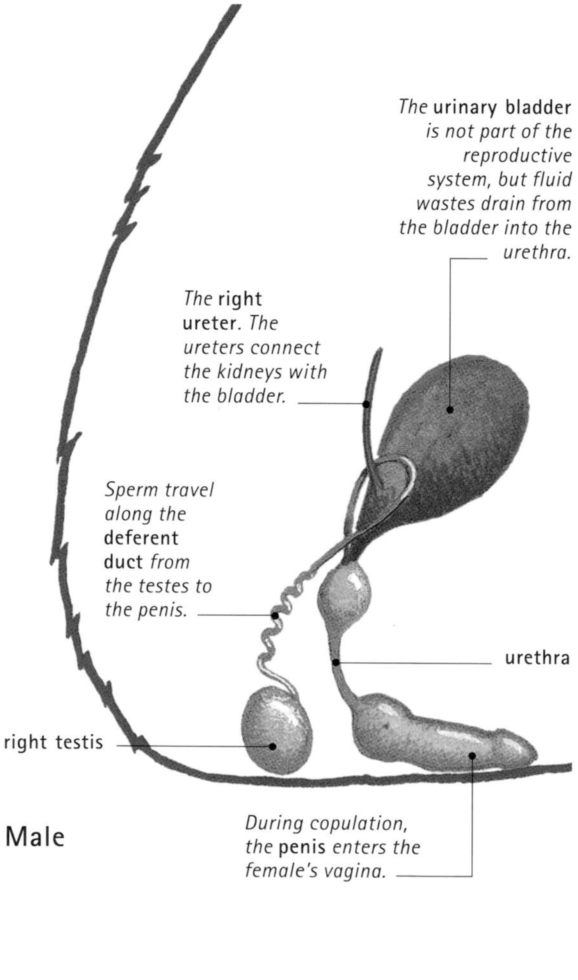

The urinary bladder is not part of the reproductive system, but fluid wastes drain from the bladder into the urethra.
The right ureter. The ureters connect the kidneys with the bladder.
Sperm travel along the deferent duct from the testes to the penis.
urethra
right testis
During copulation, the penis enters the female's vagina.

Male

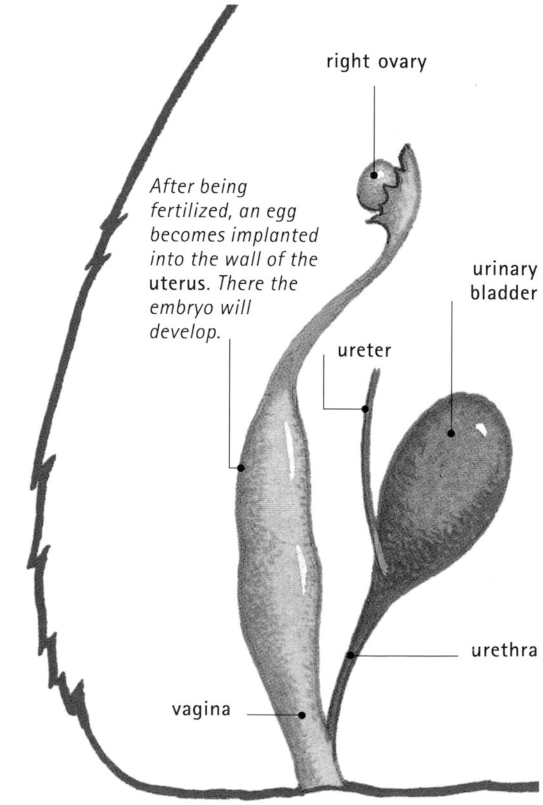

right ovary
After being fertilized, an egg becomes implanted into the wall of the uterus. There the embryo will develop.
urinary bladder
ureter
urethra
vagina

Female

▶ Having spent the first three months of their lives in their mother's den, bear cubs are playful and inquisitive. The cubs grow rapidly during their first year. This is due to the rich nature of bear milk, which contains more than 20 percent fat.

(340–860 g), less than 1 percent of their mother's weight. When compared with the size they will become as adults, newborn grizzly bear cubs are smaller than the young of any other placental mammal.

A fat supply

Why are the cubs born so helpless in the middle of winter, when their mother is fast asleep? The answer lies in their need to put on as much weight as possible so they, too, can hibernate when the next winter comes. Mammal fetuses cannot receive much fat through the placenta. Only when they have been born do the bear cubs get the fats they need from their mother's milk. The female grizzly produces the milk from her store of fat. A nursing mother generally uses up to twice as much of her fat reserves during winter than a male or non-nursing mother. She will typically lose up to 40 percent of her body weight by the time spring arrives.

Unlike most hibernating mammals, grizzly bears do not lower their body temperature much while they are dormant. This may be because they need to be ready to defend themselves if attacked by wolves or another bear and cannot wait to warm up first. However, the main reason is probably to provide a heat source for newborn cubs.

Grizzly cubs begin to eat solid food at the age of 5 months, but they will stay with their mother for between 18 months and 3 years. They become sexually mature between the ages of 4 and 6, but may continue to grow until they are 11 years old. Adult males will mate with several females each year if possible. Females, on the other hand, produce litters only once every two years.

TOM JACKSON

GENETICS

Breeding problems

In some parts of their range, grizzly bears are now very rare. This is because much of their habitat has been destroyed by loggers or developers. Many are shot on sight by people who think they are dangerous or a threat to livestock—although the dangers posed by grizzlies are vastly exaggerated. The population of grizzlies in the lower 48 states of the United States was estimated at around 100,000 in the early 20th century. Just 100 years later it was closer to 1,000.

This reduction in numbers causes a problem called inbreeding depression. If a bear is lucky enough to find a mate, there is a high chance that its mate will be a close relative. Breeding with close relatives is called inbreeding. Close relatives share many genes, and inbred offspring are less able to survive than those produced by more distantly related parents. A breeding program might help increase genetic diversity, but achieving this with wild grizzly bears is an almost impossible task.

FURTHER READING AND RESEARCH

Russell, C. and M. Enns. 2003. *Grizzly Seasons: Life with the Brown Bears of Kamchatka.* Firefly Books: Richmond Hill, Ontario, Canada.

Schneider, B. 2003. *Where the Grizzly Walks: The Future of the Great Bear.* Falcon: Guilford, CT.

Index

Page numbers in **bold** refer to main articles; those in *italics* refer to picture captions, or to names that occur in family trees.